continued on back

Multivariate Descriptive

Statistical Analysis

Multivariate Descriptive Statistical Analysis

Correspondence Analysis and Related Techniques for Large Matrices

LUDOVIC LEBART
ALAIN MORINEAU
KENNETH M. WARWICK

Translated by
Elisabeth Moraillon Berry

John Wiley & Sons

New York · Chichester · Brisbane · Toronto · Singapore

Originally published as *Techniques de la Description Statistique*:
Méthodes et Logiciels pour l'Analyse des Grands Tableaux © BORDAS 1977.

Library of Congress Cataloging in Publication Data

Lebart, Ludovic.
  Multivariate descriptive statistical analysis.

  Translation of: Techniques de la description
statistique.
  Bibliography: p.
  Includes index.
  1. Multivariate analysis—Data processing.
2. Matrices—Data processing. I. Morineau, Alain,
1940–   II. Warwick, Kenneth M., 1936–
III. Title.

QA278.L4213   1984   519.5'35   83-21904
ISBN 0-471-86743-8

Printed in the United States of America

10 9 8 7 6 5 4 3 2 1

Foreword

Correspondence analysis primarily consists of techniques for displaying the rows and columns of a two-way contingency table. These techniques provide flexible and powerful tools for statistical analysis. In many problems two or three dimensions of the table are highly informative. Graphical displays play an important role in providing insight and understanding. Many of the computational techniques and mathematical structures of correspondence analysis are similar to those of canonical analysis, multidimensional scaling, principal components, and discriminant analysis. Why, then, the emphasis in this book on the generic term, correspondence analysis?

Since the 1960's a group of statisticians in France under the strong influence of Jean-Paul Benzécri has developed a style of statistical analysis. The French title associated with this school is "Analyse des données." The most prevalent generic statistical technique associated with this school is correspondence analysis. More recently in the 1980's through the influence of Edwin Diday, I. C. Lerman, and Michel Jambu, methods of cluster analysis have become associated with "Analyse des données." The group has developed a sense of style in mathematical exposition, a point of view, and attitude toward statistical analysis. The authors have wisely chosen not to formalize this point of view into a philosophy or strategy for data analysis. Indeed, this point of view is best seen in the context of case studies. In the case studies we see a strong interplay between subject matter knowledge and formal mathematical procedures.

The authors of this book provide a mathematical exposition of major tools of "Analyse des données." The book includes many case studies from which the reader can begin to infer and understand the authors' point of view toward statistics. The translation from French into English by Elisabeth Berry preserves the style and point of view of the authors Ludovic

Lebart and Alain Morineau. In addition to the exposition of techniques, computer software is described.

The software was written by Ludovic Lebart and Alain Morineau. It reflects their view of computational efficiency and ease of use for the analysis of large data sets. One of the authors, Kenneth Warwick, has made extensive use of the statistical software in his consulting work. The exposition presented here reflects some of the practical insights that were gained in the course of his consulting work.

HERMAN P. FRIEDMAN

IBM Systems Research Institute
February 1984

Preface

This book is intended for researchers, teachers, and engineers who wish to understand and use recently developed statistical techniques for the analysis of data matrices. It contains an exposition of the methodology as well as numerous applications of these techniques. A basic knowledge of linear algebra and classical statistics is assumed.

Multivariate descriptive statistical analysis (MDSA) has been applied to a rapidly expanding number of fields in recent years. It is still too early for us to present a summary of work in progress. Instead we have chosen to present those methods we have used ourselves in various research projects, for which we have developed a library of programs. The most noteworthy feature of these methods is their ability to describe and analyze large data matrices.

Although the various examples presented here are derived from socioeconomic research, they can be applied readily to problems in the social sciences, in medicine, biology, and geography.

The methods to be described here are particularly appropriate for the analysis of large data matrices. This choice is motivated partly by the authors' personal experiences and partly by our belief that there is a real need for techniques that can be used with large data sets.

Statisticians and researchers often characterize these methods as exploratory, in that their goal is to obtain a summary description of a large amount of data. Large data sets often contain so many interrelationships that it is almost impossible to interpret them at first sight.

The word "descriptive" in the book's title should not lead the reader to assume that these methods are purely descriptive in nature. Certainly description is the first step toward understanding one's data. However, these techniques also allow us to analyze, verify, test, and prove certain hypotheses. Calling these methods descriptive is intended to emphasize that, in many cases, we are dealing with a large undifferentiated set of data, rather than the more precise data collected when using an experimental design. In addition, it implies that we are less involved with building a specific statistical model, or with the procedures of statistical inference.

Chapters I and II are concerned with descriptive principal components analysis and with two-way correspondence analysis, respectively. The techniques can be viewed as two different forms of principal axes analysis. Both methods attempt to represent a rectangular data matrix in terms of a reduced number of dimensions; both employ the singular value decomposition algorithm to create this reduced space; both simultaneously provide a reduced space and a graphical display of both the rows and columns of the original matrix. They differ with respect to the type of data to be analyzed. Descriptive principal components analysis should be used when the data is measured on a continuous or interval scale. Two-way correspondence analysis is appropriate for categorical or nominal data.

Chapter III provides a brief review of the classical multivariate techniques of canonical analysis and multiple discriminant analysis. The goal of this chapter is to relate these classical techniques to multiple correspondence analysis.

In Chapter IV multiple correspondence analysis is shown to be a generalization of two-way correspondence analysis. Several practical applications of this technique are presented.

Chapter V discusses clustering techniques that can be applied to very large data sets. These techniques can be employed in two ways. First, as data reduction techniques to simplify subsequent analysis; second, as an aid for the exploration of the space spanned by the first axes derived from any principal axes analysis.

Chapter VI is a purely technical chapter that discusses the computational algorithms developed to handle very large data matrices.

In Chapter VII various methods for evaluating the reliability of these techniques are discussed. The extent to which inferential statistics can be applied to the results of these techniques is considered, and various methodological problems that can occur in applying these techniques are discussed.

In Chapter VIII we have provided a FORTRAN listing of an easy-to-use version of correspondence analysis, together with a test example and output. This simplified program has been taken from a much larger system of programs called SPAD, which contains computer implementations of all the multivariate descriptive statistical analysis methods described in this book.

In publishing an English edition, we have attempted to integrate all of the recent work in this area. The last chapter of the French edition, written by N. Tabard, which described some specific socioeconomic applications of these methods, is not present in this edition. Instead we have developed new examples that have been integrated into the new appropriate chapters.

The work published here was done at CREDOC and CEPREMAP. The authors would like to thank the directors of these institutes, A. Babeau and C. Fourgeaud, who provided the facilities to carry out this work. In particular, the authors would like to thank Professor Jean-Paul Benzécri, who has inspired and stimulated most of the work described here; without him the book would not have been possible. We would also like to thank the director of research at CNRS, Edmond Lisle, Edmond Malinvaud of INSEE, and Herman Friedman of IBM Systems Research Institute, who has provided much critical insight.

Finally, we would like to thank Beatrice Shube of John Wiley & Sons, whose high standards have immeasurably improved the quality of the book.

LUDOVIC LEBART
ALAIN MORINEAU
KENNETH WARWICK

Paris
Paris
New York

February 1984

Contents

Multivariate Descriptive

Statistical Analysis

CHAPTER I

Descriptive Principal Components Analysis and Singular Value Decomposition

I.1. INTRODUCTION

There are several different ways of looking at principal components analysis. The classical statistician considers principal components analysis to be the determination of the major axes of an ellipsoid derived from a multivariate normal distribution; the axes are estimated from the sample. It was in this context that Harold Hotelling (1933) originally described principal components analysis. Most texts on classical multivariate analysis usually present the topic in this fashion (see Anderson, 1958; Kendall and Stuart, 1968; Dempster, 1969; Kshirsagar, 1972).

Psychologists, or more precisely, psychometricians consider principal components analysis as one specific type of factor analysis. Factor analysis, in its many forms, makes a variety of assumptions for dealing with communalities (specific item variance) and error variance. A detailed account of modern psychometric thinking about factor analysis can be found in Horst (1965), Lawley and Maxwell (1970), and Mulaik (1972), Harman (1976), and Cattell (1978).

More recently, data analysts have taken an entirely different point of view about principal components analysis. They have used principal components analysis without making any assumptions about distributions or an underlying statistical model. Rather, they have used it as a technique for describing a set of data in which certain algebraic or geometric criteria are

1

optimized. The availability of high speed digital computers has accelerated the growth of this approach, although it is, in fact, the original method outlined by Karl Pearson (1901). One way of looking at principal components analysis is implicit in Pearson's original formulation, but Rao's (1964) review article presents the topic in a way most closely allied to the philosophy described in this book.

In succeeding chapters we present a variety of related methods for analyzing a data matrix, which can be loosely described under the heading of correspondence analysis. In this introductory chapter we develop the theoretical basis common to all of these methods. This stems from the work of Eckart and Young (1936) on the singular value decomposition of a matrix. We emphasize the geometrical aspects of this technique.

I.1.1. Scope of the Method

The data analyst is frequently faced with the task of analyzing a rectangular matrix in which the columns represent variables (e.g., anthropometric measurements, psychological test scores, or economic indicators) and the rows represent measurements of specific objects or individuals on these variables.

For example, in biometry it is not unusual to collect a variety of measurements on a group of animals, or a sample of organs. Similarly, in economics, data may be collected on a variety of measures related to household expenditures or corporate activity. This data, when displayed in tabular form, is often very unwieldy and difficult to comprehend. The problem is that the information is not readily absorbed because of its sheer volume.

The methods to be described in this book are designed to summarize the information contained in tables like those described above and simultaneously to provide the analyst with a clear visual or geometric representation of the information.

Principal components analysis in the context of this book differs from correspondence analysis (see Chapter II) in one important respect. In principal components analysis the columns of the data matrix are generally a set of variables or measurements, whereas the rows are a relatively homogeneous sample of objects or individuals on which measurements have been made. Correspondence analysis, on the other hand, treats the rows and columns of the data matrix in a symmetric fashion.

Principal components analysis can be considered a theoretically appropriate method for the analysis of data that derives from a multivariate normal distribution. Correspondence analysis, in contrast, is theoretically the more appropriate method for the analysis of data in the form of a contingency table. In succeeding chapters we discuss the analysis of data matrices that do not precisely meet these requirements.

In all of the examples described in this chapter the rows of the tables are treated as specific observations and the columns are called variables.

Throughout the book we use the following notation. The data matrix is denoted by the letter **R**. This matrix **R** is of the order (n, p) (i.e., it has n rows and p columns). The symbol r_{ij} will designate the ith row of the jth variable. In this chapter, n is considered as the number of rows, or number of observations, and p is considered as the number of variables. The transpose of **R** is **R**′ and it has p rows and n columns.

Matrices are represented using bold capital letters; vectors are described using bold lowercase letters. Other notation is defined as it is needed.

I.1.2. A Geometric Interpretation of Principal Components Analysis

In order to understand how principal components analysis can be used as a data reduction technique, it is useful to consider geometrically the rows and columns of the matrix **R** as points in a space of p or n dimensions.

The n rows are treated as n points in a p-dimensional space called $\mathbb{R}^p$, and the p columns are treated as p points in an n-dimensional space called $\mathbb{R}^n$.

(a)   The n observations in $\mathbb{R}^p$.   For the analyst the distances between the points that represent specific observations have an obvious meaning. Two points that are close in the space $\mathbb{R}^p$ must have similar values on the p variables (e.g., two households that are close to one another have similar patterns of expenditures).

(b)   The p variables in $\mathbb{R}^n$.   In this case if two variables appear close together in the space, it implies that they have similar values over the sample of observations. In other words, they are measuring almost the same thing.

However this is a very crude interpretation of distances between points in either space. We have not considered the problem of scales of measurement. For example, how can we compute the distance between two variables if one is measured in inches and the other in pounds? How do we interpret an average distance in the variable space $\mathbb{R}^p$? When we see that two observations are close together in the variable space, does this mean that they have similar values on all of the p variables, or on just some of them? Principal components analysis provides us with an answer to these questions.

Let us assume initially that the distances between the variables and between the observations have some substantive meaning. Principal components analysis enables us to represent the points graphically after a series of iterations in which we attempt to find a reduced number of dimensions (or a subspace) that provides a good fit for the observations and the variables so that the distances between the points in the subspaces provide an accurate representation of the distances in the original data matrix.

The application of principal components analysis to a data matrix involves two kinds of considerations: the first is purely mathematical, and the second, statistical. We make an effort to distinguish these two issues so that the potential user can gain a clear understanding of the mathematical basis of the technique, as opposed to statistical issues that require the judgment and knowledge of the analyst.

I.2.   GENERAL ANALYSIS

Let X be a matrix with n rows and p columns. For example, the matrix X can have 2000 rows and 300 columns, which represent the values of 300 variables on 2000 observations.

There are 600,000 cells in the matrix X. We assume that there is a pattern of meaningful structural relationships among the 300 variables. The problem to be faced is, can we "summarize" these 600,000 data points into a considerably smaller number of linear combinations of the original data points without losing a great deal of information and without distorting the relationships between the original variables?

What we are looking for is a data reduction technique that can be applied in a systematic fashion to a variety of data matrices. This technique should provide a rapid approximation to the original data matrix. We describe this data reduction technique by looking at the vector spaces $\mathbb{R}^p$ and $\mathbb{R}^n$, using the example in which $p = 300$ and $n = 2000$.

I.2.1.   Fitting the Data Points in $\mathbb{R}^p$

Each of the n rows of matrix X can be considered either a vector or a point in $\mathbb{R}^p$. The entire matrix constitutes a set of n points in $\mathbb{R}^p$. The summarization problem is solved by finding a q-dimensional subspace of $\mathbb{R}^p$ such that q is considerably smaller than p and such that the n points are approximated in this subspace.

Let us assume, for example, that the 2000 row points are contained in a 10-dimensional subspace. Or, more accurately, let us say that the n points' true locations can be recovered in a satisfactory manner from their positions in this subspace. Then only two kinds of information are needed in order to find the n points' relative positions in $\mathbb{R}^p$: the new basis (for example, 10 300-dimensional vectors) and the points' new coordinates in this basis (for example, 2000 10-dimensional vectors).

The 600,000 original numbers can therefore be recovered from the 23,000 numbers defined above ($10 \times 300 + 2000 \times 10 = 23,000$).

We begin by finding a one-dimensional vector subspace, that is, a straight line passing through the origin, which represents the best possible fit of the set of points. This is illustrated in Figure 1.

Let **u** be a unit vector. We also denote its transpose by **u'**. The relationship **u'u** = 1 means that **u** is a unit vector.

The projection OH_i of a vector OM_i on the one-dimensional subspace defined by **u** is none other than the scalar product of OM_i and **u**, that is, the sum of the one to one products of the components of OM_i and **u**.

Each one of the n rows of matrix **X** is a vector in $\mathbb{R}^p$. The product matrix of matrix **X** and vector **u** is a column matrix with n elements, each of which is the scalar product of one row of **X** and **u**.

Thus the n components of the column matrix **Xu** are also the projections of the n points of the data set onto the vector **u**.

Among the criteria used for fitting a set of n points to a subspace, the classical least squares method is undoubtedly the most widespread method, and computationally the easiest. Least squares consists of minimizing the sum of squares of the distances:

$$\sum_{i=1}^{n} M_i H_i^2 \qquad (1)$$

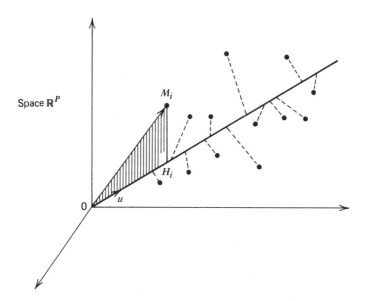

Figure 1

Applying the Pythagorean theorem to each of the n right triangles H_iOM_i gives us the following relationship (the summations are performed on the n values of the index i):

$$\sum M_iH_i^2 = \sum OM_i^2 - \sum OH_i^2 \tag{2}$$

Given that the sum $\sum OM_i^2$ is a characteristic of the set of points and therefore a fixed sum, the following maximization is equivalent to the minimization of the above quantity:

$$\sum OH_i^2 \tag{3}$$

This quantity is expressed as a function of $\mathbf{X}$ and $\mathbf{u}$ by

$$\sum OH_i^2 = (\mathbf{Xu})'\mathbf{Xu} = \mathbf{u}'\mathbf{X}'\mathbf{Xu} \tag{4}$$

To determine $\mathbf{u}$, we must therefore find the maximum of the quadratic form $\mathbf{u}'\mathbf{X}'\mathbf{Xu}$, under the constraint $\mathbf{u}'\mathbf{u} = 1$.

Let $\mathbf{u}_1$ be the vector for which this maximum is reached.

The two-dimensional subspace that is the best fit for the set of points obviously contains the subspace defined by $\mathbf{u}_1$. (If the two-dimensional space did not contain $\mathbf{u}_1$, there would exist a better subspace containing $\mathbf{u}_1$.) Therefore we can find $\mathbf{u}_2$, the second vector in the basis for this subspace: $\mathbf{u}_2$ is orthogonal to $\mathbf{u}_1$ and maximizes $\mathbf{u}_2'\mathbf{X}'\mathbf{Xu}_2$.

The q-dimensional subspace, with $q \leq p$, that is best in the least squares sense, is found in similar fashion. The solution for this maximization problem is shown in Section I.5.

The results are as follows: $\mathbf{u}_1, \mathbf{u}_2, \ldots, \mathbf{u}_q$ are the orthogonal eigenvectors of the matrix $\mathbf{X}'\mathbf{X}$ corresponding to the q largest eigenvalues, ranked in descending order:

$$\lambda_1 \geq \lambda_2 \geq \cdots \geq \lambda_q$$

I.2.2. Relationship Between Fit in $\mathbb{R}^p$ and $\mathbb{R}^n$

These results show that the unit vector $\mathbf{u}_1$ that characterizes the one-dimensional subspace that is the best fit for the n individual points in $\mathbb{R}^p$ is the eigenvector of $\mathbf{X}'\mathbf{X}$ that corresponds to the largest eigenvalue λ_1.

More generally, the q-dimensional subspace that is the best fit, in the least squares sense, of the points in $\mathbb{R}^p$ is generated by the q first eigenvectors of the symmetric matrix $\mathbf{X}'\mathbf{X}$ corresponding to the q largest eigenvalues.

Now, let us look at $\mathbb{R}^n$, where matrix $\mathbf{X}$ can be represented by p variable points whose n coordinates are the columns of $\mathbf{X}$.

The problem of finding a unit vector $\mathbf{v}$ (and subsequently a q-dimensional subspace) that is the best fit for the points in $\mathbb{R}^n$ leads us to maximize the sum of squares of the p projections on $\mathbf{v}$, those projections being the p components of the vector $\mathbf{X}'\mathbf{v}$.

The quantity to be maximized is

$$(\mathbf{X}'\mathbf{v})'\mathbf{X}'\mathbf{v} = \mathbf{v}'\mathbf{X}\mathbf{X}'\mathbf{v} \tag{5}$$

with the constraint $\mathbf{v}'\mathbf{v} = 1$.

We retain, as previously, the q eigenvectors of $\mathbf{X}\mathbf{X}'$ corresponding to the q largest eigenvalues. Let $\mathbf{v}_\alpha$ be the eigenvector of $\mathbf{X}\mathbf{X}'$ that corresponds to the eigenvalue μ_α.

In $\mathbb{R}^p$ we had the equation

$$\mathbf{X}'\mathbf{X}\mathbf{u}_1 = \lambda_1\mathbf{u}_1 \tag{6}$$

In $\mathbb{R}^n$ we now have

$$\mathbf{X}\mathbf{X}'\mathbf{v}_1 = \mu_1\mathbf{v}_1 \tag{7}$$

We premultiply the two elements of equation (6) by $\mathbf{X}$. We thus obtain

$$(\mathbf{X}\mathbf{X}')\mathbf{X}\mathbf{u}_1 = \lambda_1(\mathbf{X}\mathbf{u}_1) \tag{8}$$

This equation shows that, for any eigenvector $\mathbf{u}_1$ of $\mathbf{X}'\mathbf{X}$ relative to a nonzero eigenvalue λ_1, there exists an eigenvector $\mathbf{X}\mathbf{u}_1$ of $\mathbf{X}\mathbf{X}'$, relative to the same eigenvalue λ_1.

Recalling that we called μ_1 the largest eigenvalue of $\mathbf{X}\mathbf{X}'$, the relationship $\lambda_1 \leq \mu_1$ is necessarily true.

By premultiplying the two elements of equation (7) by $\mathbf{X}'$, we see that $\mathbf{X}'\mathbf{v}_1$ is an eigenvector of $\mathbf{X}'\mathbf{X}$ relative to the eigenvalue μ_1, which leads us to the relationship $\mu_1 \leq \lambda_1$; this finally proves that $\lambda_1 = \mu_1$.

We could demonstrate in a similar fashion that this series of the nonzero eigenvalues of the two matrices $\mathbf{X}'\mathbf{X}$ and $\mathbf{X}\mathbf{X}'$ are identical.

Therefore, it is unnecessary to repeat the computations for diagonalizing $\mathbf{X}\mathbf{X}'$, since by a simple linear transformation associated with the original matrix $\mathbf{X}$, we obtain the vectors $\mathbf{X}\mathbf{u}_\alpha$ in $\mathbb{R}^n$.

Note that the norm of the vector $\mathbf{X}\mathbf{u}_\alpha$ is λ_α (since $\mathbf{u}'_\alpha\mathbf{X}'\mathbf{X}\mathbf{u}_\alpha = \lambda_\alpha$) and therefore the unit vector $\mathbf{v}_\alpha$ corresponding to the same eigenvalue λ_α is given, for $\lambda_\alpha \neq 0$, by the relationship

$$\mathbf{v}_\alpha = \frac{1}{\sqrt{\lambda_\alpha}}\mathbf{X}\mathbf{u}_\alpha \tag{9}$$

We find, in symmetric fashion, for every α such that $\lambda_\alpha \neq 0$;

$$\mathbf{u}_\alpha = \frac{1}{\sqrt{\lambda_\alpha}} \mathbf{X}' \mathbf{v}_\alpha \tag{10}$$

$\mathbf{u}_\alpha$ is called the αth principal axis in $\mathbb{R}^p$. $\mathbf{v}_\alpha$ is called the αth principal axis in $\mathbb{R}^n$.

Note. In the subspace of $\mathbb{R}^p$ generated by $\mathbf{u}_\alpha$, the coordinates of the points are the components of $\mathbf{X}\mathbf{u}_\alpha$; they are also the components of $\mathbf{v}_\alpha\sqrt{\lambda_\alpha}$.

The coordinates of the points on a principal axis in $\mathbb{R}^p$ are therefore proportional to the components of the principal axis in $\mathbb{R}^n$ that corresponds to the same eigenvalue (and vice versa).

I.2.3. Recreating the Original Data

We always call $\mathbf{u}_\alpha$ the αth eigenvector of the matrix $\mathbf{X}'\mathbf{X}$, which has norm 1, and which corresponds to the eigenvalue λ_α; and we call $\mathbf{v}_\alpha$ the αth eigenvector of norm 1 of $\mathbf{X}\mathbf{X}'$. Equation (11) is written, for the αth axes of $\mathbb{R}^p$ and $\mathbb{R}^n$;

$$\mathbf{X}\mathbf{u}_\alpha = \sqrt{\lambda_\alpha}\,\mathbf{v}_\alpha \tag{11}$$

We postmultiply the two elements of this equation by $\mathbf{u}'_\alpha$, and sum over all of the axes (some of which may correspond to a zero eigenvalue; they complete the orthonormal basis formed by the preceding axes):

$$\mathbf{X}\sum_{\alpha=1}^{p} \mathbf{u}_\alpha\mathbf{u}'_\alpha = \sum_{\alpha=1}^{p} \sqrt{\lambda_\alpha}\,\mathbf{v}_\alpha\mathbf{u}'_\alpha \tag{12}$$

Let us call $\mathbf{U}$ the matrix of order (p, p) whose columns are the eigenvectors $\mathbf{u}_\alpha$ of $\mathbf{X}'\mathbf{X}$. Since these vectors are orthogonal and of norm 1, we have $\mathbf{U}'\mathbf{U} - \mathbf{I}$ (identity matrix), and therefore $\mathbf{U}\mathbf{U}' = \mathbf{I}$.

However,

$$\sum_{\alpha=1}^{p} \mathbf{u}_\alpha\mathbf{u}'_\alpha = \mathbf{U}\mathbf{U}' \tag{13}$$

Since the eigenvalues λ_α are always ranked in descending order, the above formula becomes

$$\mathbf{X} = \sum_{\alpha=1}^{p} \sqrt{\lambda_\alpha}\,\mathbf{v}_\alpha\mathbf{u}'_\alpha \tag{14}$$

This formula may be considered a formula for recreating matrix $\mathbf{X}$ on the basis of the λ_α eigenvalues and their associated eigenvectors $\mathbf{u}_\alpha$ and $\mathbf{v}_\alpha$. This presentation of the general analysis is not unlike the work of numerical analysts on singular value decomposition (cf. Sylvester, 1889, for square matrices, and Eckart and Young, 1936, 1939, for rectangular matrices).

If the $p - q$ smallest eigenvalues are "very small," the summation can be limited to the first q terms:

$$\mathbf{X} \simeq \mathbf{X}^* = \sum_{\alpha=1}^{q} \sqrt{\lambda_\alpha}\, \mathbf{v}_\alpha \mathbf{u}'_\alpha \tag{15}$$

If q is much smaller than p, it is easy to see the advantage by comparing the two elements of this equation: vector $\sqrt{\lambda_\alpha}\, \mathbf{v}_\alpha$ has n components and vector $\mathbf{u}_\alpha$ has p components. The np terms of $\mathbf{X}$ are therefore approximated by terms constructed from the $q(n + p)$ values contained on the right side of the equation.

The recovery of the original data can be evaluated by the quantity

$$\tau_q = \frac{\sum\limits_{i,j} x_{ij}^{*2}}{\sum\limits_{i,j} x_{ij}^2} \tag{16}$$

Again, we have

$$\tau_q = \frac{\operatorname{tr} \mathbf{X}^{*\prime}\mathbf{X}^*}{\operatorname{tr} \mathbf{X}'\mathbf{X}} \tag{17}$$

(tr is the trace). By substituting $\mathbf{X}$ and $\mathbf{X}^*$ with their values in equations (14) and (15), we obtain

$$\tau_q = \frac{\sum\limits_{\alpha \leq q} \lambda_\alpha}{\sum\limits_{\alpha=1}^{p} \lambda_\alpha} \tag{18}$$

The quantity τ_q, which is less than or equal to 1, is called the percentage of variance relative to the q first factors. Its interpretation as a measure of the numerical quality of the recovery is clear, but we see later that the problem of its statistical significance is a sensitive issue.

I.3.   APPLYING PRINCIPAL COMPONENTS ANALYSIS

We now show how to adapt principal components analysis to a situation where the problem is to *describe* rather than to simply *summarize* large data sets.

Let us consider matrix **R** whose columns are 300 responses or measurements taken on 2000 individuals. The order of the matrix is then: $n = 2000$, $p = 300$. Specifically, let us assume that the data consist of values of 300 types of annual expenditures made by 2000 individuals.

We would like to understand how the 300 expenditures are related to one another as well as whether similarities exist among the behavior patterns of the individuals.

I.3.1.   Analysis in $\mathbb{R}^{P}$

In the variable space we attempt to fit the set of n points, first in a one-dimensional, then in a two-dimensional subspace so as to obtain, on a graphical display, the most faithful possible visual representation of the distances between the individuals with respect to their p expenditures.

(a)   The problem is no longer one of maximizing the sums of squares of the projections of the points' distances from the origin. What now has to be maximized is the sum of squares of the distances between *all the pairs of individuals*. In other words, the best fitting line H_1 is no longer limited to passing through the origin as was H_0 in the previous discussion (see Figure 2).

Let h_i and h_j represent the values of the projections of two individual points i and j on H_1. We have the following equation:

$$\sum_{i,j}^{n} (h_i - h_j)^2 = n\sum_i h_i^2 + n\sum_j h_j^2 - 2\left(\sum_i h_i\right)\left(\sum_j h_j\right) \tag{19}$$

$$= 2n^2\left(\frac{1}{n}\sum_i h_i^2 - \bar{h}^2\right) = 2n\sum_i (h_i - \bar{h})^2 \tag{20}$$

where $\sum_{i,j}^{n}$ is a summation for all i's and j's less than or equal to n. $\bar{h}$ is the mean of the projections of the n individuals, and is therefore also the projection of the *centroid* G of the set of points onto H_1.

We note that the point with abscissa $\bar{h}$ on H_1 is the projection of point G whose jth coordinates are

$$\bar{r}_j = \frac{1}{n}\sum_i r_{ij} \tag{21}$$

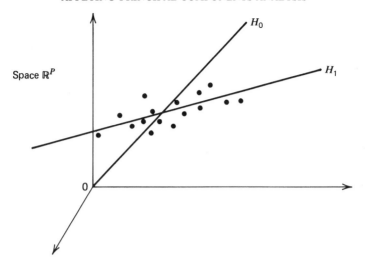

Figure 2

Thus if the origin is placed in G, the quantity to be maximized becomes again the sum of squares of the distances to the origin, which brings us back to the problem discussed above (Section I.2) in $\mathbb{R}^P$.

The required subspace is obtained by first transforming the data matrix into a matrix $\mathbf{X}$ whose general term is $x_{ij} = r_{ij} - \bar{r}_j$, and then performing the analysis on $\mathbf{X}$.

(b)   The distance between two individuals k and k' is written, in $\mathbb{R}^P$,

$$d^2(k, k') = \sum_{j=1}^{p} (r_{kj} - r_{k'j})^2 \qquad (22)$$

In this summation there might be some values of j for which the corresponding variables are of very different orders of magnitude: for example expenditures on stamps, expenditures on rent. It may be necessary in some cases, particularly when the units of measurement are different, to give each variable the same weight in defining the distances among individuals. Normalized principal axes analysis is used in this situation. The measurement scales are standardized by using the following distance measure:

$$d^2(k, k') = \sum_{j=1}^{p} \left(\frac{r_{kj} - r_{k'j}}{s_j\sqrt{n}} \right)^2 \qquad (23)$$

s_j is the standard deviation of variable j:

$$s_j^2 = \frac{1}{n} \sum_{i=1}^{n} (r_{ij} - \bar{r}_j)^2 \tag{24}$$

Finally, we note that the normalized analysis in $\mathbb{R}^p$ of the raw data matrix $\mathbf{R}$ is also the general analysis of $\mathbf{X}$, whose general term is

$$x_{ij} = \frac{r_{ij} - \bar{r}_j}{s_j \sqrt{n}} \tag{25}$$

In this space, we must therefore diagonalize the matrix $\mathbf{C} = \mathbf{X}'\mathbf{X}$. The general term $c_{jj'}$ of this matrix is written

$$c_{jj'} = \sum_i x_{ij} x_{ij'} \tag{26}$$

That is,

$$c_{jj'} = \frac{(1/n) \sum_i (r_{ij} - \bar{r}_j)(r_{ij'} - \bar{r}_{j'})}{s_j s_{j'}} \tag{27}$$

$c_{jj'}$ is the correlation coefficient between variables j and j'. (This is why $\sqrt{n}$ is introduced into the denominator of equation (25).)

The coordinates of the n individual points on the principal axis $\mathbf{u}_\alpha$ (the αth eigenvector of matrix $\mathbf{C}$) are the n components of the vector $\hat{\mathbf{v}}_\alpha = \mathbf{X}\mathbf{u}_\alpha$.

The abscissa of the individual point i on this axis is explicitly written as

$$\hat{v}_{\alpha i} = \sum_{j=1}^{p} u_{\alpha j} x_{ij} = \sum_{j=1}^{p} u_{\alpha j} \frac{r_{ij} - \bar{r}_j}{s_j \sqrt{n}} \tag{28}$$

I.3.2. Analysis in $\mathbb{R}^n$

The general analysis developed in Section I.2 shows that fitting a set of points in one space implies fitting the other set in the other space as well. The purpose of transforming the initial matrix $\mathbf{R}$, using equation (25), was twofold: first, to obtain a fit that accounts in the best possible way for the distances between individual points; second, to give equal weights to each of the variables in defining the distances among the individuals.

We note that equation (25) does not treat the rows and columns of the initial matrix $\mathbf{R}$ in a symmetric fashion.

What is the meaning of the distance between two variables j and j' in $\mathbb{R}^n$, if the columns of the transformed matrix $\mathbf{X}$ are used as coordinates for these variables?

Let us calculate the Euclidean distance between two variables j and j':

$$d^2(j, j') = \sum_{i=1}^{n} (x_{ij} - x_{ij'})^2 \tag{29}$$

$$d^2(j, j') = \sum_{i=1}^{n} x_{ij}^2 + \sum_{i=1}^{n} x_{ij'}^2 - 2 \sum_{i=1}^{n} x_{ij} x_{ij'} \tag{30}$$

When we substitute for x_{ij} its value taken from equation (25), taking into account the relationship

$$s_j^2 = \frac{1}{n} \sum_{i=1}^{n} (r_{ij} - \bar{r}_j)^2 \tag{31}$$

we see that

$$\sum_{i=1}^{n} x_{ij}^2 = \sum_{i=1}^{n} x_{ij'}^2 = 1 \tag{32}$$

(All the variable points are located on a sphere of radius 1, whose center is at the origin of the axes.) We also see that

$$\sum_{i=1}^{n} x_{ij} x_{ij'} = c_{jj'} \tag{33}$$

We obtain the relationship of the distance between two variable points j and j' and these variables' correlation coefficient $c_{jj'}$:

$$d^2(j, j') = 2(1 - c_{jj'}) \tag{34}$$

The distances between variables generated by equation (25) have the following characteristics:

1. Two variables that are highly correlated are located either very near one another ($c_{jj'} = 1.0$) or as far away as possible from one another ($c_{jj'} = -1.0$), depending on whether they are positively or negatively correlated.

2. Two variables that are orthogonal (uncorrelated) ($c_{jj'} = 0$) are at a moderate distance from one another.

Note that the analysis in $\mathbb{R}^n$ is not performed relative to the centroid of the variable points.

We have seen that it is unnecessary to diagonalize matrix $\mathbf{XX}'$, of order (n, n) once the eigenvalues λ_α and the eigenvectors $\mathbf{u}_\alpha$ of matrix $\mathbf{C} = \mathbf{X}'\mathbf{X}$ are known. The reason is that the vector

$$\mathbf{v}_\alpha = \frac{1}{\sqrt{\lambda_\alpha}} \mathbf{Xu}_\alpha \qquad (35)$$

is the unit eigenvector of $\mathbf{XX}'$ relative to the eigenvalue λ_α. The abscissas of the variable points on the axis are the components of $\mathbf{X}'\mathbf{v}_\alpha$, that is, of $\mathbf{u}_\alpha\sqrt{\lambda_\alpha}$, and are by-products of the computations that have already been performed in the other space.

Note.   The cosine of two variable vectors in space $\mathbb{R}^n$ is the correlation coefficient between the two variables. If the two variables are located at a distance of 1 from the origin (i.e., if they have unit variance), the cosine is their scalar product.

Thus the p correlation coefficients that exist among the coordinates of the points on an axis and the p variables are the p components of the vector:

$$\mathbf{X}'\mathbf{v}_\alpha = \frac{1}{\sqrt{\lambda_\alpha}} \mathbf{X}'\mathbf{Xu}_\alpha = u_\alpha\sqrt{\lambda_\alpha} \qquad (36)$$

The abscissa of a variable point on an axis is the correlation coefficient between this variable and a *created variable* (the linear combination of the initial variables) that constitutes the axis.

It is important to note that the interpretation of the transformation represented by equation (25) differs greatly in $\mathbb{R}^n$ and in $\mathbb{R}^p$. Consider, for instance, the operation of centering the variables:

1. In $\mathbb{R}^p$ the transformation $r_{ij} \rightarrow (r_{ij} - \bar{r}_j)$ is equivalent to translating the axes' origin to the center of gravity (or centroid) of the points.
2. In $\mathbb{R}^n$ the transformation $r_{ij} \rightarrow (r_{ij} - \bar{r}_j)$ is a projection parallel to the first bisector of the axes. (The general term of the (n, n) matrix $\mathbf{P}$ that is associated with this transformation is
    $p_{ii'} = \delta_{ii'} - 1/n$ where $\delta_{ii'} = 1$ if $i = i'$, and $\delta_{ii'} = 0$ otherwise.)

I.3.3.   Supplementary Variables and Supplementary Individuals

It often happens, in practice, that additional information is available that might be added to matrix $\mathbf{R}$. We might have some data on the n individuals that could be added to the p variables being analyzed but that is of a somewhat different nature than the p variables. For example, we might wish to add income level and age of the individuals to a data set consisting of food consumption variables.

On the other hand, we may have data on the p variables for a number of additional individuals who belong to a control group, and who therefore cannot be included in the sample.

Matrix $\mathbf{R}$ thus has additional rows and columns:

1. A matrix $\mathbf{R}^+$ with n rows and p_s columns added to its columns.

2. A matrix $\mathbf{R}_+$ with n_s rows and p columns added to its rows.

The matrix $\mathbf{R}_+^+$, in which both individuals and variables are supplementary, is unnecessary for the analysis; see Figure 3.

Matrices $\mathbf{R}^+$ and $\mathbf{R}_+$ are transformed, respectively, into matrices $\mathbf{X}^+$ and $\mathbf{X}_+$ in order to make the new rows and columns comparable to those of $\mathbf{X}$.

(a) In $\mathbb{R}^n$, we have p_s supplementary variable points. In order to remain consistent in interpreting intervariable distances in terms of correlations, the following transformation must be performed (normalized principal axes analysis):

$$x_{ij}^+ = \frac{r_{ij}^+ - \bar{r}_j^+}{s_j^+ \sqrt{n}} \qquad (37)$$

We compute *new means* and new *standard deviations* that incorporate the supplementary variables, in order to place these supplementary variables on the sphere of unit radius.

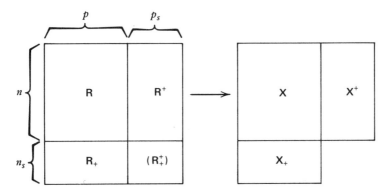

Figure 3

The *projection operator* on the axis of $\mathbb{R}^n$ is the unit vector

$$\mathbf{v}_\alpha = \frac{1}{\sqrt{\lambda_\alpha}}\mathbf{Xu}_\alpha \tag{38}$$

The abscissas of the p_s supplementary variables on this axis are therefore the p_s components of the vector $\mathbf{X}^{+\prime}\mathbf{v}_\alpha$.

(b)   In $\mathbb{R}^p$, placement of the supplementary individuals in relation to the others consists of positioning them relative to the centroid of the points (which has already been calculated) and then dividing the coordinates by the standard deviations of the variables (which have already been calculated for the n individuals). Therefore the following transformation is performed:

$$x_{+ij} = \frac{r_{+ij} - \bar{r}_j}{s_j\sqrt{n}} \tag{39}$$

The projection operator on axis α of $\mathbb{R}^p$ is the unit vector $\mathbf{u}_\alpha$. The abscissas of the n_s supplementary individuals are therefore the n_s components of the vector $\mathbf{X}_+\mathbf{u}_\alpha$.

Let $\mathbf{X}_s$ be the matrix

$$\begin{bmatrix} \mathbf{X} \\ \mathbf{X}_+ \end{bmatrix}$$

The product $\mathbf{X}_s\mathbf{u}_\alpha$ simultaneously yields the $n + n_s$ coordinates of the original individuals plus the supplementary individuals.

The variables or individuals involved directly in fitting the data are sometimes called *active* elements (active variables or active individuals). The supplementary variables are called *illustrative* variables.

I.3.4.   Nonparametric Analysis

The only difference between nonparametric methods and what we have already described is that they require a preliminary transformation of the data. These techniques should be used when the data is heterogeneous. They provide extremely robust results, and also lend themselves to statistical interpretation.

(a)   Analysis of Ranks.   Let us transform the original data matrix into a matrix of ranks. In this matrix an observation i on a variable j is q_{ij}. q_{ij} is the rank of observation i after the n observations have been ranked. Under

these circumstances the distance between two variables j and j' is defined by the formula

$$d^2(j, j') = \frac{6}{n(n-1)(n+1)} \sum_i (q_{ij} - q_{ij'})^2 \qquad (40)$$

The reader will recognize that $(1 - d^2(j, j'))$ is Spearman's rank-order correlation coefficient.

Rank-order analysis is best applied in the following situations:

(1)   The basic data consists of ranks, in which case there is no choice.

(2)   The variables' scales of measurement are so different from one another that principal components data reduction operations remain inadequate. However, this procedure is not a remedy for asymmetric distributions. Finally, analyzing a set of ranks is easier to justify than analyzing an extremely heterogeneous set of measurements.

(3)   The implicit assumptions applied to the measurements are weaker, and consequently less arbitrary. The distribution of the distances is now nonparametric; confidence levels, as in any nonparametric test, are dependent only upon the hypothesis that the observations are distributed continuously, which is more plausible in this context than the assumption of normality.

(4)   Finally, this method is robust, in the sense that it is very insensitive to the presence of outliers, often an appreciable advantage.

The results are interpreted in the same way as a regular principal components analysis. The reason is that principal components analysis is performed after transformation into ranks. Note that standardizing the variables is unnecessary here, since all of the ranks have the same variance.

The distance between two variables is interpreted in terms of *rank correlation*: two variables are close to one another if their ranks are similar across all the observations. Conversely, two variables are distant from one another if their ranks are almost totally opposite. Two observations are close when their ranks are similar for each of the variables. When the variables and observations are mapped simultaneously, the relative location of a variable vis-à-vis all of the observations gives an idea of the total configuration of the observations' rank for this variable.

Finally, the nonparametric nature of the results allows us to perform *tests* of validity on the eigenvalues. This is because the rules governing the eigenvalues of a matrix of ranks depend only on the parameters n and p, the number of rows and columns of the matrix. It is therefore possible to construct tables to determine the significance levels of the eigenvalues (cf. section VII.3.1).

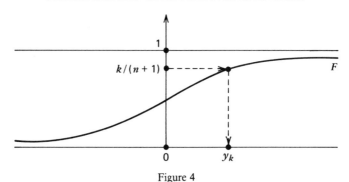

Figure 4

(b) Robust Axes Analysis. The least squares criterion is best applied to a normal distribution. When the distribution is uniform, excessive weight is given to observations at both extremes. In this case, the analysis becomes more *robust* if we apply a transformation that "normalizes" the uniform distribution of the ranks.

Let us consider the kth observation among the n ranked observations. Let F be the distribution function for a normal distribution. We substitute the rank observation k with the value y_k obtained from the *inverse distribution function* of the normal distribution:

$$y_k = F^{-1}\left(\frac{k}{n+1}\right) \tag{41}$$

This type of transformation can be found in Fisher and Yates (1949); see Figure 4.

For a large n, the transformation is equivalent to replacing the kth observation with the expected value of the kth observation in a ranked sample of n normal observations.

1.4.   IMPLEMENTATION OF PRINCIPAL COMPONENTS ANALYSIS

I.4.1.   Presentation and Interpretation of Results

It is useful to present the configurations obtained in each of the two spaces on one map. In doing so, some rules of interpretation based upon the preceding theoretical considerations must be kept in mind. The ideal way to present the two configurations is to superimpose one as a transparency on top of the other. In practice they are presented most often on one map using

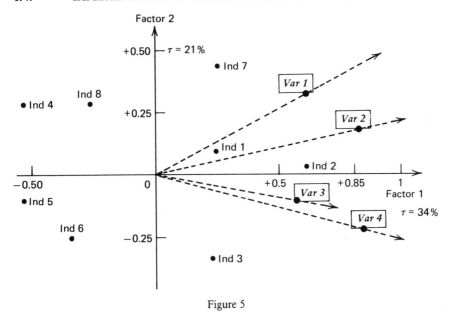

Figure 5

two different types of symbols. Figure 5 shows an example of a map for $p = 4$ and $n = 8$ (four variables and eight individuals).

Note that the individual points are distributed evenly around the origin, their centroid. On the other hand, all of the variable points may well be situated to one side of one of the axes, since the analysis of the p variable points in $\mathbb{R}^n$ is done relative to the origin.

Recall that the cosine of the angle between two variable points in $\mathbb{R}^n$ is the correlation coefficient between these two variables. The better the fit, the better this relationship will be maintained in the projection on the map.

Distances between individual points are interpreted in terms of similar patterns of response to the variables. Distances between variables are interpreted (when using normalized principal components analysis), in terms of correlation. However, a great deal of caution is needed in interpreting the distance between a variable point and an individual point, because these two points do not belong to the same space.

However, we can legitimately compare the relative positions of two individuals with respect to the entire set of variables, or of two variables with respect to the entire set of individuals. The dotted line axes of Figure 5 are scaled projections of the axes of $\mathbb{R}^p$ (each corresponding to one variable) onto the plane of the first two factors, that is, the plane that is the best fit in this space of the set of n individual points.

Let $\mathbf{u}_\alpha$ be the unit vector in $\mathbb{R}^p$ representing the factor axis α and let $\mathbf{e}_j$ be the unit vector corresponding to the initial axis whose coordinates correspond to the jth variable (the kth component of $\mathbf{e}_j$ is equal to δ_{kj}). The scalar product of $\mathbf{u}_\alpha$ and $\mathbf{e}_j$ is

$$\mathbf{u}'_\alpha \mathbf{e}_j = \sum_{k=1}^{p} \delta_{kj} u_{\alpha k} = u_{\alpha j} \tag{42}$$

But the coordinate of the jth variable on axis α is equal to $u_{\alpha j}\sqrt{\lambda_\alpha}$.

Therefore the dotted lines on Figure 5 show a perspective of the original system of axes, while taking into account relationships that exist among the initial variables. On the map, for example, individuals 4 and 5 display similar patterns of behavior characterized by low values on all four variables, whereas individual 2 has a higher score for these variables. Thus keeping these precautions in mind, *simultaneous representation* of the two sets of points allows us to analyze distances within one data matrix, and provides a description of the elements responsible for these distances.

The percentages shown on the axes define their explanatory power: they represent the proportion of total variance that is explained by each axis. Thus the value of 34% associated with the first axis indicates that the first eigenvalue represents 34% of the total variance (i.e., the trace of the matrix that has been diagonalized).

This is however an extremely conservative measure of the explanatory power of the axes, which is sometimes linked rather arbitrarily to the scaling of the data. We discuss this subject further in Chapter VII.

I.4.2. Example

(a) The Data Set. In a socioeconomic survey about the living conditions and aspirations of the French, 1085 employed individuals were asked in 1978 to give their own estimates of the average monthly salary or income for eight occupations:

1. Unskilled laborer (UNSK).
2. Foreman (FORE).
3. Engineer (ENGI).
4. Chief executive officer (CEOF).
5. Post office clerk (CLER).
6. High school teacher (TEAC).
7. Small store owner (STOR).
8. Physician (PHYS).

We wish to describe the relationships between the income estimates and the characteristics of the respondents.

The eight estimates thus constitute the set of active variables. Two other continuous variables were available for the same individuals: actual age (AGE) and age on leaving school or college (AGE*). These two variables are projected afterwards, as supplementary elements (see Section I.3.3).

All that was known about the respondents was their answers to the following six questions (nominal variables):

1. Sex (two categories).
2. Marital status (five categories).
3. Age and sex (combined) (eight categories).
4. Ownership of stocks and shares, (four categories).
5. Interest in the survey (estimated by the surveyor: four categories).
6. Opinion about the following statement: "The family is the only place where we feel good."

As a matter of fact, the survey contained a great deal more information (the example of Chapters IV and V deals with the same original data set). In a real investigation, it is interesting to use the largest possible set of characteristics in order to discover some unexpected relationships.

Table 1 shows the means and standard deviations of the eight estimates (in 1978 dollars). We must keep in mind that we are dealing with perceived incomes and not with actual incomes.

The correlation matrix (Table 2) shows that all the coefficients are positive. The standard deviation of a correlation coefficient under the hypothesis of independence is approximately 0.03 (which is $1/\sqrt{n-1}$ for an n of 1085); a range of ± 0.06 thus gives the approximate 95% confidence intervals for the correlation coefficient.

Table 1.

Identifiers	Mean	Standard Deviations
UNSK	325.96	49.89
FORE	620.07	148.71
ENGI	1241.24	429.41
CEOF	3431.00	1894.48
CLER	442.68	88.20
TEAC	715.44	280.05
STOR	1267.94	675.35
PHYS	3062.82	1496.26

Table 2. Matrix of Correlations

	UNSK	FORE	ENGI	CEOF	CLER	TEAC	STOR	PHYS
UNSK	1.00	.33	.02	.03	.21	.06	.03	.02
FORE	.33	1.00	.31	.14	.22	.07	.06	.18
ENGI	.02	.31	1.00	.22	.09	.11	.07	.12
CEOF	.03	.14	.22	1.00	.03	.19	.08	.26
CLER	.21	.22	.09	.03	1.00	.18	.03	.04
TEAC	.06	.07	.11	.19	.18	1.00	.08	.18
STOR	.03	.06	.07	.08	.03	.08	1.00	.24
PHYS	.02	.18	.12	.26	.04	.18	.24	1.00

The two first axes account for 40.4% of the variance (Table 3). This percentage should not be interpreted as a percentage of information, since the variance corresponding to the remaining axes may be purely random noise (see Chapter VII).

In Figure 6 one can see that all the points representing the estimates are on the left side of the vertical axis containing the origin. The first factor is consequently a "size factor," expressing the fact that the main variability between the individuals consists of the general level of the values: that is to say, people tend to give high estimates across occupations, or low estimates.

The second axis describes differences in the social status of the various professions.

We note that the age of the respondent (AGE) is slightly correlated with the general level of the estimates. Conversely, age when leaving school or college (AGE*) seems to be associated with low values.

The standard deviation of the coordinates of a supplementary variable drawn at random from a normal distribution (or derived from a random permutation of a finite set of numerical values) is $\sigma = (n - 1)^{-1/2}$ (here:

Table 3.

Eigenvalues	Percentage	Cumulated Percentage
1.942	24.27	24.27
1.291	16.14	40.41
1.026	12.82	53.23
0.977	12.21	65.44
0.829	10.36	75.80
0.718	8.98	84.78
0.693	8.67	93.45
0.524	6.55	100.00
Total 8.000		

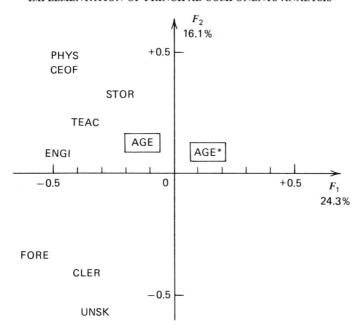

Figure 6. Principal component analysis of eight estimates.

$2\sigma \cong 0.06$, as shown previously), since this coordinate is simply the correlation coefficient between two independent random variables. Thus the two supplementary points have significant abscissas on the horizontal axis (see Table 4).

(b) Characteristics of the Individuals. It is pointless to present the scattergram of the 1085 individuals. We can, however, plot the centroids of the various categories that are the responses to the six questions.

Figure 7 shows a graphical display of the various categories: the estimates of the various incomes are clearly related to several categories.

Age and sex play a predominant role. Men give higher estimates than women, and young people of both sexes seem to underestimate the incomes of all the occupations. The cross-tabulated variable age–sex shows a divergence between older men and women. However, there are only 20 employed women of more than 60 years in the sample. Consequently, we must not give too much importance to this last phenomenon.

As shown in Chapter IV (Section IV.4.5), it is possible to assess the significance of each category, taking into account its frequency and coordinates. We assign to each point the value of a variable that has a standardized normal distribution (test-value).

Table 4.

Variables	Axis 1	Axis 2	Axis 3
Active variables			
UNSK	−0.40	−0.62	−0.21
FORE	−0.66	−0.38	0.23
ENGI	−0.54	0.07	0.60
CEOF	−0.52	0.40	0.29
CLER	−0.43	−0.46	−0.13
TEAC	−0.46	0.19	−0.29
STOR	−0.33	0.37	−0.51
PHYS	−0.54	0.46	−0.20
Supplementary variables			
AGE	−0.15	0.02	−0.02
AGE*	0.11	0.07	0.03

Table 5 gives the coordinates of some centroids on the two first axes, and the corresponding "test-values."

The points "living together" or "widow," for instance, are not significantly correlated with the first two axes, since the corresponding "test-values" V_1 and V_2 are within the range $[-2, +2]$. All the points in the list are dependent on at least one axis.

We can check that, despite their low frequency, women over 60 years have a significant coordinate on the vertical axis.

Figure 7 shows that ownership of stocks and shares, conservative opinions about family life, and interest in the survey are also correlated with the level of estimate of all these incomes and salaries.

It would be dangerous to interpret these relationships in terms of causality. We confine ourselves to pointing out the two main steps of such analysis:

1. Displaying the individuals on the basis of level and structure of a set of numerical variables (Figure 6 shows the pattern of variables).

2. Identifying them afterwards using all available characteristics (Figure 7).

In so doing, the levels of the estimates of the incomes corresponding to the various occupations seem to be related to social status, degree of social integration, age, and sex of the respondents.

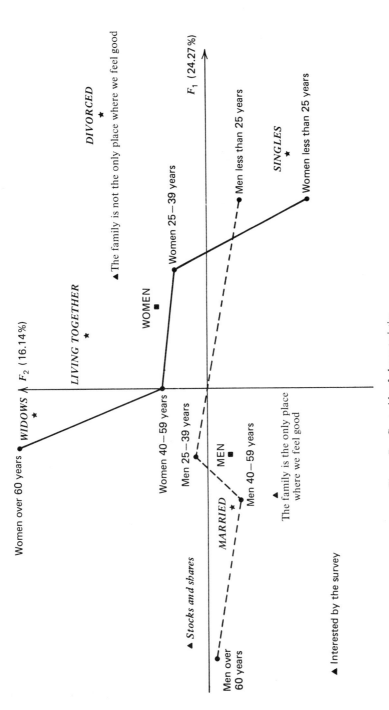

Figure 7. Centroids of characteristics.

25

Table 5.   *Coordinates and test-values of some categories*

		Coordinates		Test-values	
	Numbers	F_1	F_2	V_1	V_2
Sex					
Male	667	-0.04	-0.01	-3.07	-0.87
Female	418	0.06	0.01	3.07	0.87
Marital status					
Single	217	0.12	-0.02	3.96	-1.02
Married	738	-0.05	-0.01	-4.61	-1.02
Living together	52	0.05	0.08	0.78	1.42
Divorced	46	0.16	0.07	2.26	1.24
Widow	32	-0.04	0.13	-0.45	1.80
Ownership of stocks and shares					
Yes	72	-0.9	0.01	-3.43	0.18
Age and sex					
Men $+60$ years	32	-0.31	-0.01	-3.65	-0.10
Women $+60$ years	20	-0.06	0.24	-0.54	2.65

I.5.   MATHEMATICAL APPENDIX

Maximization of a Quadratic Form Under a Quadratic Constraint

Let us find a vector **u** that maximizes the quantity **u′Au**, with the constraint **u′Mu** = 1 where **A** and **M** are symmetric matrices; in addition, **M** is positive definite.

The problem here is more general than the problem discussed earlier (see Section I.2.1), where **A** = **X′X** and **M** = **I**. This formulation proves useful for correspondence analysis and discriminant analysis. Fortunately it does not complicate the proofs that follow.

We give two elementary proofs for solving this problem. One proof uses Lagrange multipliers (in the classical derivation of constrained maxima or minima). The other proof assumes that certain spectral properties of the symmetric matrices are known.

(a)   Direct Proof (Lagrange Multipliers).   The quadratic form **u′Au** is written

$$\mathbf{u'Au} = \sum_i \sum_j a_{ij} u_i u_j \qquad (43)$$

When we derive this quantity for the p components of vector **u** successively, we see that the vector of the partial derivatives of **u′Au** is written in matrix form:

$$\frac{\partial(\mathbf{u'Au})}{\partial \mathbf{u}} = 2\mathbf{Au} \tag{44}$$

and

$$\frac{\partial \mathbf{u'Mu}}{\partial \mathbf{u}} = 2\mathbf{Mu} \tag{45}$$

Finding a maximum requires setting the derivatives of the Lagrangian equation equal to 0:

$$\mathscr{L} = \mathbf{u'Au} - \lambda(\mathbf{u'Mu} - 1) \tag{46}$$

where λ is a Lagrange multiplier. Consequently, the equation

$$\frac{\partial \mathscr{L}}{\partial \mathbf{u}} = 2\mathbf{Au} - 2\lambda \mathbf{Mu} = 0 \tag{47}$$

expresses the necessary condition for a maximum or minimum.

From this we deduce the equation

$$\mathbf{Au} = \lambda \mathbf{Mu} \tag{48}$$

When we premultiply the two elements of this equation by $\mathbf{u'}$, taking into account the fact that $\mathbf{u'Mu} = 1$, we obtain

$$\lambda = \mathbf{u'Au} \tag{49}$$

The value of the parameter λ is the maximum.

When matrix $\mathbf{M}$ is positive definite, therefore nonsingular, the relationship is written

$$\mathbf{M^{-1}Au} = \lambda \mathbf{u} \tag{50}$$

$\mathbf{u}$ is the eigenvector of the matrix $\mathbf{M^{-1}A}$ that corresponds to the largest *eigenvalue* λ (if it is unique, as is generally the case).

Henceforth we call $\mathbf{u}_1$ the vector $\mathbf{u}$ that corresponds to the largest value λ_1 such that equation (48) is true. Let us find the vector $\mathbf{u}_2$ that is $\mathbf{M}$-orthogonal to $\mathbf{u}_1$ (such that $\mathbf{u'_1Mu_2} = 0$), and unitary (such that $\mathbf{u'_2Mu_2} = 1$), and which maximizes the quadratic form $\mathbf{u'_2Au_2}$.

We set equal to zero the partial derivatives

$$\mathscr{L} = \mathbf{u}_2'\mathbf{A}\mathbf{u}_2 - \lambda_2\left(\mathbf{u}_2'\mathbf{M}\mathbf{u}_2 - 1\right) - \mu_2\mathbf{u}_2'\mathbf{M}\mathbf{u}_1 \tag{51}$$

where λ_2 and μ_2 are two Lagrange multipliers.

The condition for a maximum or minimum is written, for $\mathbf{u}_2$,

$$\frac{\partial \mathscr{L}}{\partial \mathbf{u}_2} = 2\mathbf{A}\mathbf{u}_2 - 2\lambda_2\mathbf{M}\mathbf{u}_2 - \mu_2\mathbf{M}\mathbf{u}_1 = 0 \tag{52}$$

If we multiply the elements of this equation by $\mathbf{u}_1'$, we see that $\mu_2 = 0$, since

$$\mathbf{u}_1'\mathbf{A}\mathbf{u}_2 = \lambda_1\mathbf{u}_1'\mathbf{M}\mathbf{u}_2 = 0 \tag{53}$$

This leaves

$$\mathbf{A}\mathbf{u}_2 = \lambda_2\mathbf{M}\mathbf{u}_2 \tag{54}$$

When $\mathbf{M}$ is nonsingular, $\mathbf{u}_2$ is the second eigenvector of $\mathbf{M}^{-1}\mathbf{A}$ relative to the second largest eigenvalue λ_2, if it is unique. This proof is easily extended to the case of a unit vector $\mathbf{u}_\alpha$ (i.e., $\mathbf{u}_\alpha'\mathbf{M}\mathbf{u}_\alpha = 1$), which is $\mathbf{M}$-orthogonal to the $\mathbf{u}_\beta$ vectors found above ($\mathbf{u}_\alpha'\mathbf{M}\mathbf{u}_\beta = 0$, for $\beta < \alpha$) and which maximizes the form $\mathbf{u}_\alpha'\mathbf{A}\mathbf{u}_\alpha$. Then we have

$$\mathbf{A}\mathbf{u}_\alpha = \lambda_\alpha\mathbf{M}\mathbf{u}_\alpha \tag{55}$$

And if $\mathbf{M}$ is nonsingular,

$$\mathbf{M}^{-1}\mathbf{A}\mathbf{u}_\alpha = \lambda_\alpha\mathbf{u}_\alpha \tag{56}$$

Of course, α cannot become greater than p, the order of matrix $\mathbf{A}$.

(b) Second Proof. We limit ourselves to outlining this proof for the case where $\mathbf{M}$ is positive definite. Then we can decompose this matrix in its classical form: $\mathbf{M} = \mathbf{L}'\mathbf{L}$, where $\mathbf{L}$ is nonsingular because $\mathbf{M}$ is assumed to be positive definite.

By setting $\mathbf{u} = \mathbf{L}^{-1}\mathbf{y}$, the constraint for standardization $\mathbf{u}'\mathbf{M}\mathbf{u} = 1$ is now written $\mathbf{y}'\mathbf{y} = 1$, and the quantity to be maximized, $\mathbf{u}'\mathbf{A}\mathbf{u}$, becomes

$$\mathbf{y}'\mathbf{A}_1\mathbf{y}, \quad \text{with } \mathbf{A}_1 = \mathbf{L}'^{-1}\mathbf{A}\mathbf{L}^{-1} \tag{57}$$

The symmetric matrix $\mathbf{A}_1$ can be diagonalized. Let $\mathbf{T}$ be the orthogonal matrix (p, p) whose columns are the eigenvectors $\mathbf{t}_\alpha$ of $\mathbf{A}_1$ after standardization, and arranged in order of decreasing value of the eigenvalues λ_α. Let Λ be the diagonal matrix whose αth element has the value λ_α.

Let us set $\mathbf{z} = \mathbf{T}'\mathbf{y}$ (which implies $\mathbf{y} = \mathbf{T}\mathbf{z}$ because $\mathbf{T}' = \mathbf{T}^{-1}$). We then have

$$\mathbf{y}'\mathbf{A}_1\mathbf{y} = \mathbf{y}'\mathbf{T}\Lambda\mathbf{T}'\mathbf{y} = \mathbf{z}'\Lambda\mathbf{z} \tag{58}$$

with the constraint $\mathbf{z}'\mathbf{z} = 1$.

We note that

$$\lambda_1 \geq \mathbf{z}'\Lambda\mathbf{z} \tag{59}$$

This is because

$$\lambda_1 - \mathbf{z}'\Lambda\mathbf{z} = \mathbf{z}'(\lambda_1\mathbf{I} - \Lambda)\mathbf{z} \geq 0 \tag{60}$$

The maximum λ_1 is reached for $\mathbf{z}' = (1, 0, 0, 0, \ldots, 0)$, therefore, for $\mathbf{y} = \mathbf{t}_1$ and thus for $\mathbf{u}_1 = \mathbf{L}^{-1}\mathbf{t}_1$.

From the equation

$$\mathbf{A}_1\mathbf{t}_1 = \lambda_1\mathbf{t}_1 \tag{61}$$

we obtain

$$\mathbf{L}'^{-1}\mathbf{A}\mathbf{L}^{-1}\mathbf{t}_1 = \lambda_1\mathbf{t}_1 \tag{62}$$

Then

$$\mathbf{L}'^{-1}\mathbf{A}\mathbf{L}^{-1}\mathbf{t}_1 = \lambda_1\mathbf{L}\mathbf{u}_1 \tag{63}$$

and, finally,

$$\mathbf{M}^{-1}\mathbf{A}\mathbf{u}_1 = \lambda_1\mathbf{u}_1 \tag{64}$$

We note in passing that it is sufficient here to diagonalize a *symmetric* matrix $\mathbf{A}_1$ (after decomposing $\mathbf{M}$ in the form $\mathbf{M} = \mathbf{L}'\mathbf{L}$), whereas the preceding matrix, $\mathbf{M}^{-1}\mathbf{A}$, is nonsymmetric. This property is used in computing algorithms (particularly in correspondence analysis) because finding spectral elements is faster and more efficient for symmetric matrices.

CHAPTER II

Correspondence Analysis

Like principal components analysis, correspondence analysis can be presented in several different ways. In fact, it is difficult to trace the method's history accurately (see, e.g., Hill, 1974; Benzecri, 1976; Nishisato, 1980). The underlying theory probably dates back to Fisher's (1940) work on contingency tables. The context of this first presentation was strictly classical inferential statistics. However, since Benzecri's (1964, 1973) work, the emphasis has been mostly on the algebraic and geometric properties of the descriptive tool provided by the method. This analysis is not simply a special case of principal components analysis, although it is possible to use that technique by performing appropriate transformations on the variables (provided each space is treated separately). Correspondence analysis can be viewed as finding the best *simultaneous representation* of two data sets that comprise the rows and columns of a data matrix (cf. Section II.2.5).

Later we shall outline the various mathematical bases of the method.

Correspondence analysis and principal components analysis are used under different circumstances. Principal components analysis is used for tables consisting of continuous measurements. Correspondence analysis is best applied to contingency tables (cross-tabulations). By extension, it also provides a satisfactory description of tables with binary coding (cf. Chapter IV).

We use a concrete example: a contingency table cross-tabulating answers to two questions concerning jobs in a survey:

1. Type of work activity (26 categories).
2. The job's main advantage, as stated by the individual (17 categories).

This example is described in greater detail in Section II.4. The contingency table is shown in Table 1.

We see how the structure of the data itself dictates the ways in which the configuration of points is constructed, the distance is chosen, and a goodness of fit criterion is selected.

II.1. GEOMETRY OF THE CONFIGURATION OF POINTS AND GOODNESS OF FIT CRITERION

II.1.1. Construction of the Configuration

First, let us consider the space $\mathbb{R}^p$, that is, $\mathbb{R}^{17}$, where the matrix is represented by 26 activity-points.

The data matrix, **K**, has n rows and p columns; k_{ij} represents the number of individuals belonging to activity i and having stated advantage j.

If the components of the vectors of $\mathbb{R}^p$ are made up of raw frequencies such as k_{ij}, the distances that exist among the activity-points will not provide meaningful interpretations. There are work activities where the frequencies are large numbers, and for which the 17 advantages are mentioned often, whereas other work activities have small frequencies, and therefore the advantages necessarily appear less frequently, and are located near the origin. Thus we are not interested in analyzing raw frequencies, but rather in distribution profiles, that is the distributions of the advantages within each type of activity; in other words, the conditional frequencies of stating advantage j, given that an individual works in activity i.

Similarly, in space $\mathbb{R}^n$, that is $\mathbb{R}^{26}$, where the points are the 17 advantages, we are interested in looking at the job sector profiles of the advantages. These are the conditional frequencies of working in activity i given that advantage j has been stated.

Notation. Let

$$k_{i.} = \sum_{j=1}^{p} k_{ij} \qquad (1)$$

be the total number of persons in activity i;

$$k_{.j} = \sum_{i=1}^{n} k_{ij} \qquad (2)$$

Table 1.

Type of Work Activity	Variety of Work	Freedom	Human Contact	Schedules	Salaries	Job Security	Family Life	Interesting
Farming-fishing	4	189	0	3	2	2	9	3
Farm-food industry	1	13	3	10	17	12	4	1
Energy-mines	1	9	1	0	4	13	0	2
Steel	5	5	2	9	18	5	3	2
Chemical-glass-oil	2	7	1	4	15	5	2	1
Wood-paper	2	5	0	4	1	0	3	0
Auto-aviation-shipping	2	3	1	8	16	17	1	8
Textile-leather-shoes	3	18	0	6	16	5	4	4
Pharmaceutical-industries	3	7	3	6	6	0	0	2
Manufacturing	0	18	1	12	31	7	0	8
Construction	7	63	2	9	31	9	4	6
Food-grocery	2	43	16	7	6	4	7	1
Small business	8	95	23	15	15	2	13	7
Miscellaneous business	5	32	9	9	17	4	5	4
Administrative services	8	26	10	24	24	80	10	17
Telecommunications	1	7	2	11	3	14	2	6
Social services	4	10	10	8	2	1	6	4
Health services	3	31	16	15	11	19	5	19
Teaching-research	2	33	27	31	9	18	27	24
Transportation	2	19	2	12	12	21	0	1
Insurance-banking	8	12	4	8	13	21	2	10
Domestic workers	0	8	0	4	5	2	7	1
Other services	8	35	14	13	16	10	6	25
Printing-publishing	2	13	2	14	5	8	0	10
Private services	3	26	9	3	12	5	8	8
No answer	0	14	15	3	4	4	3	4
Total	86	741	173	248	311	288	131	178

be the total number of persons stating advantage j;

$$k = \sum_{i,j} k_{ij} \qquad (3)$$

be the sample total.

We take as the jth component of the ith vector of $\mathbb{R}^p$:

$$\left\{ \frac{k_{ij}}{k_{i.}} \right\} \qquad \text{for all} \quad j = 1, 2, \ldots, p \qquad (4)$$

Table 1.   (Continued)

			Main Advantage of Job						
Near Home	Good Atmosphere	Social Advantages	I Am my Own Boss	I Like It	Other	None	Work Outdoors	No Answer	Total
12	2	1	4	11	15	12	8	1	278
8	3	5	1	9	5	11	0	0	103
2	0	2	1	4	3	6	1	0	49
6	5	5	0	2	3	22	0	0	92
6	1	2	2	3	0	5	0	1	57
2	1	1	1	1	0	3	0	2	26
7	2	4	3	6	1	24	0	1	104
13	4	2	3	6	2	26	0	2	114
6	3	3	0	2	1	8	0	0	50
19	11	3	2	10	4	26	0	6	158
9	10	3	4	14	8	35	2	2	218
8	2	0	1	6	1	7	0	3	114
9	5	2	3	13	4	18	1	3	236
7	4	3	0	8	3	18	0	3	131
11	3	8	2	6	9	16	3	4	261
3	1	1	2	1	3	5	0	2	64
2	3	1	0	3	2	1	0	1	58
10	2	3	7	24	1	5	0	5	176
3	4	43	8	18	3	11	1	3	265
4	5	5	1	3	3	13	0	1	104
4	2	5	6	3	1	10	0	3	112
5	7	2	1	2	2	11	1	1	59
6	4	10	9	11	4	14	0	1	186
0	8	3	2	5	4	11	0	2	89
4	4	2	3	10	3	8	0	2	110
1	1	2	1	5	1	3	0	3	64
167	97	121	67	186	86	329	17	52	3278

We call the ith component of the jth vector of $\mathbb{R}^n$

$$\left\{ \frac{k_{ij}}{k_{.j}} \right\} \qquad \text{for all} \quad i = 1, 2, \ldots, n \tag{5}$$

The reader should note the first fundamental difference between correspondence analysis and principal components analysis: here the transformations performed on the raw data in the two spaces are identical (because the data sets, being placed in correspondence, play similar roles). But they are different analytically: the table of new coordinates in $\mathbb{R}^p$ is not the simple transpose of the new coordinates in $\mathbb{R}^n$ (whereas in principal components analysis, the same analytical formula is derived from very different transformations).

We note the relative frequencies as follows:

$$f_{ij} = \frac{k_{ij}}{k} \qquad \left(\sum_{i=1}^{n} \sum_{j=1}^{p} f_{ij} = 1 \right) \qquad (6)$$

and, in a similar fashion,

$$f_{i\cdot} = \sum_{j=1}^{p} f_{ij} = \frac{k_{i\cdot}}{k} \qquad \left(\sum_{i=1}^{n} f_{i\cdot} = 1 \right) \qquad (7)$$

$$f_{\cdot j} = \sum_{i=1}^{n} f_{ij} = \frac{k_{\cdot j}}{k} \qquad \left(\sum_{j=1}^{p} f_{\cdot j} = 1 \right) \qquad (8)$$

We have the following relationships:

$$\frac{f_{ij}}{f_{\cdot j}} = \frac{k_{ij}}{k_{\cdot j}} \qquad \text{and} \qquad \frac{f_{ij}}{f_{i\cdot}} = \frac{k_{ij}}{k_{i\cdot}} \qquad \text{for all } i \text{ and } j \qquad (9)$$

We deal exclusively with relative frequencies.

II.1.2. Choice of Distances

Having chosen *profiles* (which are conditional frequencies) to construct the configuration of points, and wishing to obtain somewhat invariant results, we adopt a distance that is different from the customary Euclidean distance.

The distance between two activities i and i' is given by

$$d^2(i, i') = \sum_{j=1}^{p} \frac{1}{f_{\cdot j}} \left(\frac{f_{ij}}{f_{i\cdot}} - \frac{f_{i'j}}{f_{i'\cdot}} \right)^2 \qquad (10)$$

In symmetric fashion, the distance between two advantages j and j' is written

$$d^2(j, j') = \sum_{i=1}^{n} \frac{1}{f_{i\cdot}} \left(\frac{f_{ij}}{f_{\cdot j}} - \frac{f_{ij'}}{f_{\cdot j'}} \right)^2 \qquad (11)$$

The distance thus defined is called the chi-square distance.

In fact, this distance only differs from the usual Euclidean distance in that each square is weighted by the inverse of the frequency corresponding to each term.

Essentially, the reason for choosing the chi-square distance is that it verifies the property of *distributional equivalency*, expressed as follows:

1. If two advantages having identical job sector profiles are aggregated, then the distances between activities remain unchanged.
2. If two activities having identical distribution profiles are aggregated, then the distances between advantages remain unchanged.

This property is important, because it guarantees *invariance of the results* irrespective of how the variables were originally coded.

From a strictly technical point of view, it is logical to consider two points that are overlapping in space as a single point corresponding to the totals of the two aggregated categories.

Thus the representation of the activities will only change very slightly if advantages with similar profiles are aggregated. In general, there is no loss of information when certain categories are aggregated. Conversely, there is nothing to be gained by subdividing homogeneous categories.

Let us demonstrate this property in the case of aggregating two activities i_1 and i_2 into one activity i_0, whose relative population frequency f_{i_0}. satisfies the relationship

$$f_{i_0 \cdot} = f_{i_1 \cdot} + f_{i_2 \cdot}. \tag{12}$$

In expressing the distance $d^2(j, j')$ between two advantages j and j', only two terms, called T_1 and T_2, make use of i_1 and i_2:

$$T_1 + T_2 = \frac{1}{f_{i_1 \cdot}} \left(\frac{f_{i_1 j}}{f_{\cdot j}} - \frac{f_{i_1 j'}}{f_{\cdot j'}} \right)^2 + \frac{1}{f_{i_2 \cdot}} \left(\frac{f_{i_2 j}}{f_{\cdot j}} - \frac{f_{i_2 j'}}{f_{\cdot j'}} \right)^2 \tag{13}$$

After aggregation they are replaced by T_0 such that

$$T_0 = \frac{1}{f_{i_0 \cdot}} \left(\frac{f_{i_0 j}}{f_{\cdot j}} - \frac{f_{i_0 j'}}{f_{\cdot j'}} \right)^2 \tag{14}$$

Let us show that

$$T_0 = T_1 + T_2 \tag{15}$$

T_0 is written

$$T_0 = f_{i_0 \cdot} \left(\frac{f_{i_0 j}}{f_{i_0 \cdot} . f_{\cdot j}} - \frac{f_{i_0 j'}}{f_{i_0 \cdot} . f_{\cdot j'}} \right)^2 \tag{16}$$

T_1 and T_2 are written similarly; the three quantities are thus equal, since the profiles of i_1, i_2, and i_0 are identical. Thus equation (15) results from equation (12).

II.1.3. Choice of Goodness of Fit Criterion

In constructing the configurations of points in $\mathbb{R}^p$ and $\mathbb{R}^n$, choosing the *profiles* as coordinates gives weight to every activity and every advantage equally.

In order to calculate goodness of fit, it is natural to give each point a *weight that is proportional to its frequency*, so as not to overrepresent categories with small totals, and consequently to reflect the real population distribution. This weight is used in calculating the coordinates of the center of gravity as well as in the *goodness of fit* criterion. The quantity to be maximized is a sum of squares weighted by these weights.

Point i of $\mathbb{R}^p$ thus has a weight of $f_{i\cdot}$, whereas the weight of point j of $\mathbb{R}^n$ is $f_{\cdot j}$.

II.1.4. Summary

A summary of the initial elements of correspondence analysis is shown in Table 2. To be consistent with the general analysis presented in Chapter I, we use matrix notation; however, this notation is sometimes cumbersome, since matrices that define the initial transformations, the distances, and the criteria are all diagonal matrices, and can therefore be characterized by one index.

We always designate by $\mathbf{K}$ the (n, p) matrix of the frequencies, and by $\mathbf{F}$ the matrix of relative frequencies:

$$\mathbf{F} = \frac{1}{k}\mathbf{K} \quad \left(\text{with } k = \sum_{i, j} k_{ij}\right) \tag{17}$$

Table 2.

In Space $\mathbb{R}^p$	In Space $\mathbb{R}^n$
There are n *row-points* ($i = 1, 2, \ldots, n$)	There are p *column-points* ($j = 1, 2, \ldots, p$)
The p coordinates of row-point i are	The n coordinates of point j are
$\dfrac{f_{ij}}{f_{i\cdot}}$ for $j = 1, 2, \ldots, p$	$\dfrac{f_{ij}}{f_{\cdot j}}$ for $i = 1, 2, \ldots, n$
The weight of point i is $f_{i\cdot}$	The weight of point j is $f_{\cdot j}$
The distance between points i and i' is written	The distance between points j and j' is written
$d^2(i, i') = \displaystyle\sum_{j=1}^{p} \frac{1}{f_{\cdot j}} \left(\frac{f_{ij}}{f_{i\cdot}} - \frac{f_{i'j}}{f_{i'\cdot}} \right)^2$	$d^2(j, j') = \displaystyle\sum_{i=1}^{n} \frac{1}{f_{i\cdot}} \left(\frac{f_{ij}}{f_{\cdot j}} - \frac{f_{ij'}}{f_{\cdot j'}} \right)^2$

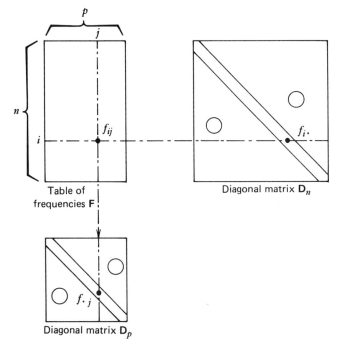

Figure 1.   Drawing of matrices $\mathbf{F}$, $\mathbf{D}_n$, and $\mathbf{D}_p$.

We designate by $\mathbf{D}_p$ the (p, p) diagonal matrix whose jth diagonal element is $f._j$; and by $\mathbf{D}_n$ the (n, n) diagonal matrix whose ith element is $f_i.$. See Figure 1. With these two matrices and matrix $\mathbf{F}$ we complete our summarization of the initial elements of the analysis, in Table 3.

II.2.   CALCULATION OF PRINCIPAL AXES AND COORDINATES

There exists a complete symmetry between the indices i and j. We limit our presentation to one space, $\mathbb{R}^p$, since the proofs in the other space are deduced by permutation of the indices i and j (that is, by transposition of $\mathbf{F}$, and substitution of the matrices $\mathbf{D}_p$ and $\mathbf{D}_n$).

We wish to represent graphically the distances between *profiles*. Therefore we place ourselves, in both spaces, at the centers of gravity of the configurations of points.

However, and this is one of the peculiarities of correspondence analysis, one can perform the analysis either relative to the *origin* or to the *center of gravity* (provided that, in the former case, the principal axis connecting the origin to the center of gravity is ignored).

Table 3.

In Space $\mathbb{R}^p$	In Space $\mathbb{R}^n$
There are n points whose *coordinates* correspond to the n rows of the matrix $\mathbf{D}_n^{-1}\mathbf{F}$	There are p points whose *coordinates* correspond to p columns of $\mathbf{F}\mathbf{D}_p^{-1}$ (or to the p rows of $\mathbf{D}_p^{-1}\mathbf{F}'$)
The *goodness of fit criterion* is characterized by the quadratic form whose matrix is $\mathbf{D}_n$	The *goodness of fit criterion* is characterized by the quadratic form whose matrix is $\mathbf{D}_p$
The *distance* is characterized by the quadratic form associated with $\mathbf{D}_p^{-1}$	The *distance* is characterized by the quadratic form associated with $\mathbf{D}_n^{-1}$

We begin with the general analysis with respect to the origin, then we show the equivalence.

II.2.1. General Analysis with Any Distance and Any Criteria

Let $\mathbf{X}$ be the (n, p) matrix of the data, $\mathbf{M}$ the symmetric positive definite matrix defining the distance in $\mathbb{R}^p$, and $\mathbf{N}$ the (n, n) diagonal matrix whose diagonal elements are the weights of the n points.

A unit vector $\mathbf{u}$ of $\mathbb{R}^p$ now satisfies the equation $\mathbf{u}'\mathbf{M}\mathbf{u} = 1$.

The n projections on axis $\mathbf{u}$ are the n rows of $\hat{\mathbf{v}} = \mathbf{X}\mathbf{M}\mathbf{u}$.

The weighted sum of squares of the projections is $\hat{\mathbf{v}}'\mathbf{N}\hat{\mathbf{v}}$, that is,

$$\mathbf{u}'\mathbf{M}\mathbf{X}'\mathbf{N}\mathbf{X}\mathbf{M}\mathbf{u}$$

This is the quantity that has to be maximized, with the constraint

$$\mathbf{u}'\mathbf{M}\mathbf{u} = 1$$

The results of Section I.5 show that $\mathbf{u}$ is the eigenvector of the matrix $\mathbf{X}'\mathbf{N}\mathbf{X}\mathbf{M}$ corresponding to the largest eigenvalue λ. Vector $\mathbf{u}$ is called a *principal axis*. It satisfies the equation $\mathbf{X}'\mathbf{N}\mathbf{X}\mathbf{M}\mathbf{u} = \lambda\mathbf{u}$.

The projection operator on axis $\mathbf{u}$, defined by $\varphi = \mathbf{M}\mathbf{u}$, is sometimes called a "*factor*." It satisfies the equation

$$\mathbf{M}\mathbf{X}'\mathbf{N}\mathbf{X}\varphi = \lambda\varphi \qquad (18)$$

The factors have a norm of 1 for the distance defined by $\mathbf{M}^{-1}$:

$$\varphi'\mathbf{M}^{-1}\varphi = \mathbf{u}'\mathbf{M}\mathbf{M}^{-1}\mathbf{M}\mathbf{u} = \mathbf{u}'\mathbf{M}\mathbf{u} = 1 \qquad (19)$$

II.2.2.   Analysis in $\mathbb{R}^p$, Calculation of Factors

In this space the n points are the n rows of $\mathbf{D}_n^{-1}\mathbf{F}$ (cf. Table 3). Let $\mathbf{u}$ be a unit vector for the distance in $\mathbb{R}^p$, that is, such that $\mathbf{u}'\mathbf{D}_p^{-1}\mathbf{u} = 1$.

The vector of the n projections on axis $\mathbf{u}$ is written $\mathbf{D}_n^{-1}\mathbf{F}\mathbf{D}_p^{-1}\mathbf{u} = \hat{\mathbf{v}}$.

The quantity to be maximized is the weighted sum of squares:

$$\hat{\mathbf{v}}'\mathbf{D}_n\hat{\mathbf{v}}$$

or

$$\mathbf{u}'\mathbf{D}_p^{-1}\mathbf{F}'\mathbf{D}_n^{-1}\mathbf{F}\mathbf{D}_p^{-1}\mathbf{u}$$

with the constraint

$$\mathbf{u}'\mathbf{D}_p^{-1}\mathbf{u} = 1$$

This problem was mentioned in the preceding paragraph; we know that $\mathbf{u}$ is the eigenvector of

$$\mathbf{S} = \mathbf{F}'\mathbf{D}_n^{-1}\mathbf{F}\mathbf{D}_p^{-1}$$

corresponding to the largest eigenvalue λ:

$$\mathbf{S}\mathbf{u} = \lambda\mathbf{u}.$$

The general term, $s_{jj'}$ of $\mathbf{S}$ is written

$$s_{jj'} = \sum_{i=1}^{n} \frac{f_{ij}f_{ij'}}{f_{i\cdot}f_{\cdot j'}} \tag{20}$$

$\mathbf{S}$ is not symmetric but we see later that the problem can be reduced to finding the eigenvectors and eigenvalues of a symmetric matrix.

Vector $\mathbf{u}$ is the first *principal axis*.

Vector $\boldsymbol{\varphi} = \mathbf{D}_p^{-1}\mathbf{u}$ is called the first *factor*.

Factor $\boldsymbol{\varphi}$ is an eigenvector of the matrix $\mathbf{A} = \mathbf{D}_p^{-1}\mathbf{F}'\mathbf{D}_n^{-1}\mathbf{F}$.

The projections of the n points on the principal axis $\mathbf{u}$ are the components of

$$\mathbf{D}_n^{-1}\mathbf{F}\mathbf{D}_p^{-1}\mathbf{u} = \mathbf{D}_n^{-1}\mathbf{F}\boldsymbol{\varphi} \tag{21}$$

More generally, if $\mathbf{u}_\alpha$ is the eigenvector of $\mathbf{S}$ corresponding to the eigenvalue λ_α, $\mathbf{u}_\alpha$ is the αth principal axis; $\boldsymbol{\varphi}_\alpha = \mathbf{D}_p^{-1}\mathbf{u}_\alpha$ is the αth factor, and the projections of the n points on axis $\mathbf{u}_\alpha$ are the components of $\mathbf{D}_n^{-1}\mathbf{F}\boldsymbol{\varphi}_\alpha$.

II.2.3. Relationship with Analysis in $\mathbb{R}^n$

Similarly, we must maximize in $\mathbb{R}^n$,

$$\mathbf{v}'\mathbf{D}_n^{-1}\mathbf{F}\mathbf{D}_p^{-1}\mathbf{F}'\mathbf{D}_n^{-1}\mathbf{v} \qquad \text{with } \mathbf{v}'\mathbf{D}_n^{-1}\mathbf{v} = 1 \tag{22}$$

The principal axis $\mathbf{v}$ is therefore the eigenvector of the matrix $\mathbf{F}\mathbf{D}_p^{-1}\mathbf{F}'\mathbf{D}_n^{-1}$ corresponding to the largest eigenvalue.

Let us rewrite the equation $\mathbf{Su} = \lambda\mathbf{u}$:

$$\mathbf{F}'\mathbf{D}_n^{-1}\mathbf{F}\mathbf{D}_p^{-1}\mathbf{u} = \lambda\mathbf{u} \tag{23}$$

Let us premultiply the two sides by $\mathbf{F}\mathbf{D}_p^{-1}$:

$$\mathbf{F}\mathbf{D}_p^{-1}\mathbf{F}'\mathbf{D}_n^{-1}\left(\mathbf{F}\mathbf{D}_p^{-1}\mathbf{u}\right) = \lambda\left(\mathbf{F}\mathbf{D}_p^{-1}\mathbf{u}\right) \tag{24}$$

Thus, as in the case of the usual distance, it appears that $\mathbf{v}$ is proportional to $\mathbf{F}\mathbf{D}_p^{-1}\mathbf{u}$. Since the $\mathbf{D}_n^{-1}$ norm of $\mathbf{F}\mathbf{D}_p^{-1}\mathbf{u}$ is equal to λ, and since we must have $\mathbf{v}'\mathbf{D}_n^{-1}\mathbf{v} = 1$, we must set

$$\mathbf{v} = \frac{1}{\sqrt{\lambda}}\mathbf{F}\mathbf{D}_p^{-1}\mathbf{u} \tag{25}$$

In analogous fashion we have

$$\mathbf{u} = \frac{1}{\sqrt{\lambda}}\mathbf{F}'\mathbf{D}_n^{-1}\mathbf{v} \tag{26}$$

The coordinates of the p variable-points on axis $\mathbf{v}$ are the components of

$$\mathbf{D}_p^{-1}\mathbf{F}'\mathbf{D}_n^{-1}\mathbf{v} = \mathbf{D}_p^{-1}\mathbf{F}'\psi \qquad \left(\text{where } \psi = \mathbf{D}_n^{-1}\mathbf{v}\right) \tag{27}$$

As before, ψ is called a *factor*, corresponding to the eigenvalue λ: it is the projection-operator on the principal axis $\mathbf{v}$. Note that

$$\psi = \mathbf{D}_n^{-1}\mathbf{v} = \frac{1}{\sqrt{\lambda}}\mathbf{D}_n^{-1}\mathbf{F}\mathbf{D}_p^{-1}\mathbf{u} = \frac{1}{\sqrt{\lambda}}\mathbf{D}_n^{-1}\mathbf{F}\varphi \tag{28}$$

And similarly

$$\varphi = \mathbf{D}_p^{-1}\mathbf{u} = \frac{1}{\sqrt{\lambda}}\mathbf{D}_p^{-1}\mathbf{F}'\mathbf{D}_n^{-1}\mathbf{v} = \frac{1}{\sqrt{\lambda}}\mathbf{D}_p^{-1}\mathbf{F}'\psi \tag{29}$$

These two sets of relationships show that *the coordinates of the points on a principal axis in a space are proportional to the components of the factor of the other space corresponding to the same eigenvalue.*

These results are immediately generalized in the case of axes corresponding to an eigenvalue λ_α: the preceding relationships still hold for φ_α, ψ_α, and λ_α.

Thus we have these relationships between *factors*:

$$\begin{cases} \psi_\alpha = \dfrac{1}{\sqrt{\lambda_\alpha}} \mathbf{D}_n^{-1} \mathbf{F}\varphi_\alpha & (30) \\[2em] \varphi_\alpha = \dfrac{1}{\sqrt{\lambda_\alpha}} \mathbf{D}_p^{-1} \mathbf{F}'\psi_\alpha & (31) \end{cases}$$

We call $\hat{\varphi}_\alpha$ and $\hat{\psi}_\alpha$ the vectors of the coordinates on the axis α. We have the relationships

$$\hat{\varphi}_\alpha = \varphi_\alpha\sqrt{\lambda_\alpha} \tag{32}$$

$$\hat{\psi}_\alpha = \psi_\alpha\sqrt{\lambda_\alpha} \tag{33}$$

Equations (30) and (31)—*transition equations*—are written explicitly for the coordinates

$$\hat{\psi}_{\alpha i} = \frac{1}{\sqrt{\lambda_\alpha}} \sum_{j=1}^{p} \frac{f_{ij}}{f_{i\cdot}} \hat{\varphi}_{\alpha j} \tag{34}$$

$$\hat{\varphi}_{\alpha j} = \frac{1}{\sqrt{\lambda_\alpha}} \sum_{i=1}^{n} \frac{f_{ij}}{f_{\cdot j}} \hat{\psi}_{\alpha i} \tag{35}$$

Thus, if we ignore the coefficient $1/\sqrt{\lambda_\alpha}$, the representative points of one configuration of points are on one axis the *barycenters* of the representative points of the configuration corresponding to the other space.

II.2.4. Supplementary Elements

Here we do not need to distinguish between supplementary variables and supplementary individuals, since rows and columns can be treated similarly.

Matrix $\mathbf{K}$ may be incremented by p_s supplementary columns (the case of n_s rows is deduced by simple permutation of indices). See Figure 2.

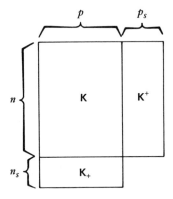

Figure 2. Drawing of supplementary rows and columns.

Again we are concerned with positioning the *profiles* of the p_s new points with respect to the p points analyzed in $\mathbb{R}^n$. Let k_{ij}^+ be the ith coordinate of the jth supplementary column. The profile of this element is the vector whose ith component is

$$\frac{k_{ij}^+}{k_{\cdot j}^+} \quad \text{with } k_{\cdot j}^+ = \sum_{i=1}^{n} k_{ij}^+ \tag{36}$$

All that remains to be done is to project this point j on axis α, using the transition formula (35):

$$\hat{\varphi}_{\alpha j}^+ = \frac{1}{\sqrt{\lambda_\alpha}} \sum_{i=1}^{n} \left(\frac{k_{ij}^+}{k_{\cdot j}^+} \right) \hat{\psi}_{\alpha i} \tag{37}$$

For a supplementary row i, we have, in analogous fashion,

$$\hat{\psi}_{\alpha i}^+ = \frac{1}{\sqrt{\lambda_\alpha}} \sum_{j=1}^{p} \left(\frac{k_{+ij}}{k_{+i\cdot}} \right) \hat{\varphi}_{\alpha j} \tag{38}$$

Like the "active" elements, the supplementary elements are calculated as pseudobarycenters.

Note. The general formula for recreating the original data (cf. Section I.2.3) may be applied here, with the following result:

$$f_{ij} = f_{i\cdot} f_{\cdot j} \left(1 + \sum_\alpha \sqrt{\lambda_\alpha}\, \psi_{i\alpha} \varphi_{j\alpha} \right) \tag{39}$$

This is sometimes called Fisher's identity (Lancaster, 1969). This equation takes into account the unit eigenvalue and its associated eigenvector (cf. Section II.5.1).

II.2.5.   Another Aspect of Correspondence Analysis—Finding the Best Simultaneous Representation

Let us consider again the (n, p) contingency table **K** cross-tabulating n types of job activities with p job advantages.

We wish to represent, on *the same axis*, the entire sets of activities and advantages so as to approximate the following ideal situation:

(a) Each advantage-point j is a *barycenter* of the activity-points, each activity i having the weight "that part of activity i that is in the profile of advantage j," in other words, the weight

$$p_i = \frac{f_{ij}}{f_{\cdot j}} \qquad \left(\sum_i p_i = 1 \right) \tag{40}$$

(b) Each activity-point i is a *barycenter* of the advantage-points, each one of these advantages having the weight "that part of advantage j in the profile of activity i," in other words, the weight

$$p'_j = \frac{f_{ij}}{f_{i\cdot}} \qquad \left(\sum_j p'_j = 1 \right) \tag{41}$$

This ideal situation is usually impossible, because it implies that each set is entirely contained in the other. (There exists a trivial solution, for which all of the points of both sets are merged with the point of abscissa 1.)

Using notation analogous to that of Section II.2.2, if φ_j designates the abscissa of the advantage j on the axis (φ_j is the jth component of a vector φ), and if ψ_i designates the abscissa of the activity i on this same axis, conditions a and b are written, respectively,

(a) $$\varphi = \mathbf{D}_p^{-1} \mathbf{F}' \psi \quad \text{and} \quad \varphi_j = \sum_{i=1}^{n} \frac{f_{ij}}{f_{\cdot j}} \psi_i \tag{42}$$

(b) $$\psi = \mathbf{D}_n^{-1} \mathbf{F} \varphi \quad \text{and} \quad \psi_i = \sum_{j=1}^{p} \frac{f_{ij}}{f_{i\cdot}} \varphi_j \tag{43}$$

These strictly barycentric equations are generally impossible to realize simultaneously. Therefore, we find the positive coefficient β that is as close as possible to 1, such that the following equations are true:

$$\varphi = \beta \mathbf{D}_p^{-1} \mathbf{F}' \psi \tag{44}$$

$$\psi = \beta \mathbf{D}_n^{-1} \mathbf{F} \varphi \tag{45}$$

Note that β is necessarily greater than or equal to 1; otherwise equations (44) and (45) would imply that each of the two sets covers an interval of the axis that is strictly contained in the interval covered by the other set. Thus we must find the smallest β such that equations (44) and (45) are satisfied.

Substituting, for example in equation (45), φ by its value taken from equation (44), we find

$$\psi = \beta^2 \mathbf{D}_n^{-1} \mathbf{F} \mathbf{D}_p^{-1} \mathbf{F}' \psi \tag{46}$$

In other words, ψ is an eigenvector of

$$\mathbf{D}_n^{-1} \mathbf{F} \mathbf{D}_p^{-1} \mathbf{F}' \tag{47}$$

relative to the largest eigenvalue

$$\lambda = \frac{1}{\beta^2} \tag{48}$$

This directly establishes that the eigenvalues λ_α derived from a correspondence analysis are all less than or equal to one.

Equations (44) and (45), where $\beta = 1/\sqrt{\lambda}$, are none other than equations (34) and (35) of Section II.2.3.

The vectors φ and ψ are the first factors corresponding to each of the sets placed in correspondence (not accounting for normalization, which is not required here).

We can extend this search for the best β-*barycentric* representation on an axis, to that of the best (β_1, β_2)-*barycentric* representation in a two-dimensional space defined by two orthogonal axes. Then we can generalize to a subspace of any dimension (less than p). Of course the representation already obtained by correspondence analysis is duplicated.

II.3. INTERPRETATION OF RESULTS

II.3.1. Introduction

As in principal components analysis, the results of a correspondence analysis are presented on *graphs* that represent the configurations of points in projection planes formed by the first principal axes taken two at a time.

A diagrammatic illustration based on the contingency table cross-tabulating types of work activities and perceived advantages is used to demonstrate the rules for interpreting a correspondence analysis.

Figure 3 shows a simplified example involving eight activities shown on the same graph with eight work advantages.

The percentages shown on the axes represent the portion of variance explained by these axes: thus F_1 (34%) means that the first eigenvalue accounts for 34% of the sum of all the eigenvalues. The first two axes together therefore explain $34\% + 20\% = 54\%$ of the total variance. This percentage gives a conservative idea of the portion of information accounted for by the principal axes. It is conservative because this is only a partial way of measuring the information; it can happen that, despite small percentages, the corresponding axes restore the bulk of the information contained in the data set (cf. Chapter VII).

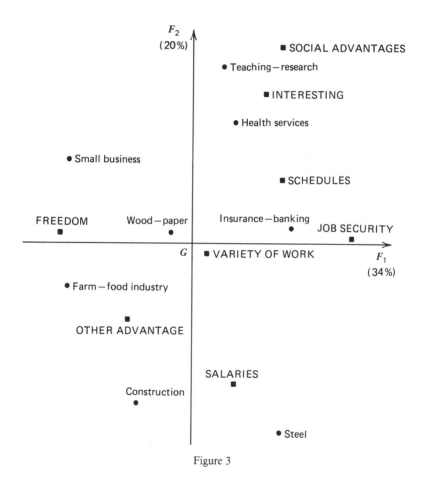

Figure 3

As in principal components analysis, it is legitimate to interpret distances among the elements of one set of points. Thus the advantages mentioned in the activities "teaching" and "health services" have similar distributions. In the other set, the presence of "social advantages" or of "interesting work" shows similar distributions among the work activities. It is also legitimate to interpret the relative positions of one point of one set with respect to *all the points* of the other set. Except in special cases, it is extremely dangerous to interpret the proximity of *two* points corresponding to different sets of points.

The center of gravity G, located at the origin of the axes, corresponds to the average profiles of both sets of points (cf. Section II.5.1). Thus the profile "variety of work," which is near the center, has an undifferentiated distribution among the activities. Similarly, the activity "wood-paper" is not differentiated among the advantages.

II.3.2. Calculation of Absolute Contributions and Squared Correlations

In order to interpret the axes, it is useful to calculate two series of coefficients for each axis (these coefficients apply equally to the rows and columns of the data matrix):

1. The *absolute contributions*, which indicate the proportion of variance explained by each variable in relation to each principal axis: this proportion is calculated with respect to the entire set of variables.

2. The *squared correlations*, which indicate the part of the variance of a variable explained by a principal axis.

(a) Absolute Contributions. The D_p^{-1}-norm of axis u_α is equal to 1: $u_\alpha' D_p^{-1} u_\alpha = 1$. Since $u_\alpha = D_p \varphi_\alpha$, it follows that the D_p-norm of factor φ_α is equal to 1: $\varphi_\alpha' D_p \varphi_\alpha = 1$, that is,

$$\sum_{j=1}^{p} f_{\cdot j} \varphi_{\alpha j}^2 = 1 \tag{49}$$

Recall that the projection of point j of $\mathbb{R}^n$ on the principal axis α is equal to

$$\sum_{i=1}^{n} \frac{f_{ij}}{f_{\cdot j}} \psi_{\alpha i} = \sqrt{\lambda_\alpha} \, \varphi_{\alpha j} = \hat{\varphi}_{\alpha j} \tag{50}$$

The variance of the set of points projected on axis α with respect to G is therefore equal to

$$\sum_j f_{\cdot j} \hat{\varphi}_{\alpha j}^2 = \sum_j f_{\cdot j} \left(\sqrt{\lambda_\alpha} \varphi_{\alpha j} \right)^2 = \lambda_\alpha \qquad (51)$$

The quotient:

$$\frac{\lambda_\alpha \varphi_{\alpha j}^2 f_{\cdot j}}{\lambda_\alpha} = f_{\cdot j} \varphi_{\alpha j}^2 = ca_\alpha(j) \qquad (52)$$

represents the absolute contribution of element j to principal axis α. Note that

$$\sum_{j=1}^{p} ca_\alpha(j) = 1 \qquad (53)$$

The absolute contribution of an element i of $\mathbb{R}^p$ to principal axis α is defined in the same way:

$$ca_\alpha(i) = f_{i\cdot} \psi_{\alpha i}^2 \qquad (54)$$

with, now,

$$\sum_{i=1}^{n} ca_\alpha(i) = 1 \qquad (55)$$

(b) Squared Correlations.   In $\mathbb{R}^n$, the square of the distance of variable j to the center of gravity is equal to (cf. Section II.5.1):

$$d^2(j, G) = \sum_{i=1}^{n} \frac{1}{f_{i\cdot}} \left(\frac{f_{ij}}{f_{\cdot j}} - f_{i\cdot} \right)^2 \qquad (56)$$

The square of the projection of variable j on axis α is equal to

$$d_\alpha^2(j, G) = \left(\sqrt{\lambda_\alpha} \varphi_{\alpha j} \right)^2 \qquad (57)$$

Note that

$$\sum_\alpha d_\alpha^2(j, G) = d^2(j, G) \qquad (58)$$

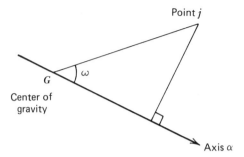

Figure 4

On Figure 4, we see that the quantity

$$\cos^2\omega = \frac{d_\alpha^2(j,G)}{d^2(j,G)} = cr_\alpha(j) \tag{59}$$

represents the part due to axis α in the position of point j. It is the *squared correlation* of axis α with element j. Note that

$$\sum_\alpha cr_\alpha(j) = 1 \tag{60}$$

What has just been said about the p elements, whose projections on the αth axis are equal to

$$\left\{ \left(\sqrt{\lambda_\alpha}\,\varphi_{\alpha j} \right); \quad j = 1,\dots,p \right\} \tag{61}$$

can be transposed to the n elements of the other set, whose projections on the αth axis are equal to

$$\left\{ \left(\sqrt{\lambda_\alpha}\,\psi_{\alpha i} \right); \quad i = 1,\dots,n \right\} \tag{62}$$

Note. In the case of normalized principal components analysis, the two concepts of absolute contributions and squared correlations coincide, with a few modifications, with those of coordinates of variable-points. In fact, the variance explained by axis α is equal to λ_α, where λ_α designates the αth eigenvalue of the correlation matrix.

But the coordinates of variable-point j on the new axes are proportional to $u_{\alpha j}\sqrt{\lambda_\alpha}$, and we have

$$\sum_j \left(u_{\alpha j}\sqrt{\lambda_\alpha}\right)^2 = \lambda_\alpha \qquad (63)$$

where $\mathbf{u}_\alpha$ is unitary. The number $100u_{\alpha j}^2$ expresses the percentage of the variance of axis α explained by variable j (absolute contribution).

On the other hand, the distance from a variable j to the origin in $\mathbb{R}^n$ is equal to 1 (cf. Section I.3.2). Thus we have, in the orthonormalized basis of the principal axis,

$$\sum_\alpha \lambda_\alpha u_{\alpha j}^2 = 1 \qquad (64)$$

The term $\lambda_\alpha u_{\alpha j}^2$ in this sum expresses the part of the total variance of variable j that is assigned to axis α (squared correlation).

(c) Example.   (The coefficient ca_α is multiplied by 100 in order to be interpreted in terms of percentages.)

Suppose that point j has an absolute contribution of 5% and a squared correlation of .75 with the first axis.

Point j thus does not have a strong influence in building this axis (if this point is far along the axis, then its weak contribution can only be due to the weakness of the weight $f_{.j}$ of variable j). On the other hand, the first axis explains 75% of variable j (its angle ω with the axis is small: $\cos^2\omega = 0.75$). This variable is almost exclusive in its characterization of the first axis.

Assume that point i has an absolute contribution of 30%, and a squared correlation of .10. This means that it contributes strongly in the creation of the first axis, but that it probably participates to the building of many other axes (the weight $f_{i.}$ and the distance to the origin $d^2(i, G)$ are both large).

II.4.   AN APPLICATION EXAMPLE

II.4.1.   Data

This study is concerned with describing a contingency table that contains *work advantages* as a function of *types of activities*.

This table was extracted from a survey conducted in France in 1979 and 1980. Among the sample's 6000 respondents, 3278 persons who were employed at the time of the survey stated what they considered to be the

Table 4. Comparison of five column-profiles with marginal profile (only a selection of rows is shown, for ease of reading; data are expressed in per thousandths [0/00])

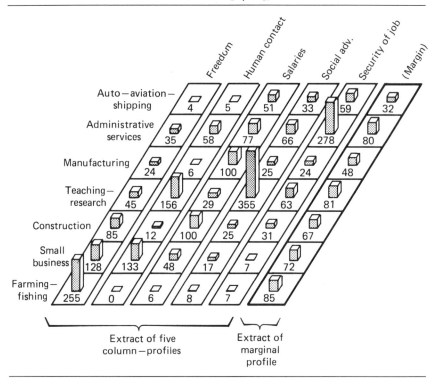

Extract of five column—profiles

Extract of marginal profile

main advantage of their job. The stated advantages were aggregated into 17 categories, which are the columns of Table 1. The respondents were grouped into 26 categories, corresponding to the rows of Table 1.

With this table, we can build row-profile and column-profile tables, which provide us, in correspondence analysis, with the row-points and column-points in the two space $\mathbb{R}^n$ and $\mathbb{R}^p$ (here, $n = 26$, $p = 17$).

For example, Table 4 shows five column-profiles, expressed as thousandths (0/00). The profiles are shown across seven rows only in order to make the comparisons of profiles easier.

Table 5 contains additional information about the rows elements (i.e., types of activities.). The first two columns of this table characterize educational level, the next three the size of the company. These five new columns allow us to define five new profile-points in $\mathbb{R}^n$; these new points are then

Table 5. Illustrative (or supplementary) variables

	Without any Diploma	High School Graduate	Size of Firm		
			-10	50 to 500	$+1000$
Farming-fishing	93	18	187	2	0
Farm-food industry	32	15	12	40	11
Energy-mines	9	8	6	17	6
Steel	29	4	4	30	27
Chemical-glass-oil	13	13	3	14	25
Wood-paper	10	5	6	11	0
Auto-aviation-shipping	25	4	7	19	61
Textile-leather-shoes	28	4	24	38	6
Pharmaceutical industries	11	11	6	16	9
Manufacturing	51	26	32	40	30
Construction	69	28	74	49	8
Food-grocery	26	9	81	13	1
Small business	34	62	163	21	3
Miscellaneous business	33	29	67	29	3
Administrative services	35	79	53	79	37
Telecommunications	4	18	10	20	10
Social services	6	25	11	10	1
Health services	17	78	32	48	29
Teaching-research	12	188	62	83	14
Transportation	23	10	20	35	14
Insurance-banking	7	44	25	24	10
Domestic workers	23	1	34	3	2
Other services	33	82	79	35	11
Printing-publishing	17	19	27	24	11
Private services	21	16	77	9	1
No answer	7	15	19	11	5
Total	668	810	1121	470	335

projected a posteriori on the graphs, using the equation for projecting supplementary points (equation (37) of Section II.2.4).

These points are in the same initial space as the "advantage-points." Their purpose is to enrich the interpretation of the graphs.

II.4.2. Results and Interpretation

The first numerical results that deserve comment are the eigenvalues and percentages of variance shown in Table 6.

It appears that three axes predominate, and account for 72.8% of the total variance. The results discussed in Chapter VII allow us to conclude that the percentage explained by the first eigenvalue is highly significant.

Table 6. Eigenvalues

SUM OF THE EIGENVALUES   0.54215372

HISTOGRAM OF THE FIRST EIGENVALUES

	EIGENVALUE	PERCENTAGE	PERCENTAGE CUM.	
1	0.19745979	36.42	36.42	**
2	0.11522371	21.25	57.67	***
3	0.08213969	15.15	72.82	***********************************
4	0.03961763	7.31	80.13	*****************
5	0.02612866	4.82	84.95	***********
6	0.01965643	3.63	88.58	********
7	0.01581012	2.92	91.49	*******
8	0.01206766	2.23	93.72	*****
9	0.00943355	1.74	95.46	****
10	0.00735602	1.36	96.82	***
11	0.00525018	0.97	97.78	**
12	0.00469712	0.87	98.65	**
13	0.00278123	0.51	99.16	*
14	0.00236718	0.44	99.60	*
15	0.00147940	0.27	99.87	*

Tables 7 and 8 show the coordinates, absolute contributions, and squared correlations of the six first principal axes for the row-points and for the column-points. The first numerical column shows the relative frequencies (masses); the second one shows the chi-square distances (dist**2) between points and the origin of the axes.

We limit ourselves here to interpreting the first two axes only.

First of all, Figure 5 shows the principal relationships among the profiles on the basis of the first two axes. We perform successive interpretations of proximities among rows (the elements of $\mathbb{R}^n$), among columns (the elements of $\mathbb{R}^p$), and finally the simultaneous representation of the two spaces.

(a)   Proximities among Rows.   (In capitals on Figure 5.) We are looking for similarities among profiles: two types of activities whose corresponding points are close to one another have similar advantages. Thus, "farming," "small business," and "food-grocery," (at the left of Figure 5) have similar advantages, and are very different from "telecommunications" or "banking" or "administrative services," which are located on the right side of the graph. Similarly, "textile," and "manufacturing" (at the bottom of Figure 5) are opposite from "teaching-research" (at the top of the graph).

The point, "wood-paper," which is near the origin (i.e., the center of gravity of the profile points), thus has a mean profile of advantages (nondiscriminating). This is only an approximation, since the coordinates of this point on axes 3 through 6 are far from negligible.

(b)   Proximities among Columns.   (In lowercase letters on Figure 5.) The stated advantages "freedom," "outdoor work" (on the left) are opposite from "security of job," whereas "human contact," "compatible with family life," and "social advantages," at the top of the graph, are opposite from "salaries," and "near home." The opposition is less pronounced along the second axis, which is not as important as the first axis.

In this case the advantage "variety of work" has a mean profile of distribution among types of activities: it is mentioned with a frequency that is just about proportional to the frequencies of the types of activities.

(c)   Proximities among Rows and Columns.   These proximities are of a different nature from the preceding ones, since the row-profiles and column-profiles are not in the same space initially. However the transition equations (equations (34) and (35)) in Section II.2.3 allow us to interpret the position of an advantage-point with respect to the entire set of activity-points, and vice versa.

Thus the point "administrative services" is near the point "insurance-banking" because their profiles of advantages are simultaneously characterized by a high percentage of the items "security of job," "schedules," and low percentages of the item "freedom."

Table 7. Coordinates and contributions of the columns

NAME	MASSES	DIST**2		COORDINATES						ABSOLUTE CONTRIBUTIONS						SQUARED CORRELATIONS					
			F1	F2	F3	F4	F5	F6	F1	F2	F3	F4	F5	F6	F1	F2	F3	F4	F5	F6
VARI	0.026	0.46	0.09	-0.05	0.02	0.14	0.02	-0.19	0.1	0.1	0.0	1.2	0.0	4.9	0.02	0.01	0.00	0.04	0.00	0.08
FREE	0.226	0.54	-0.72	-0.01	-0.12	-0.01	0.02	0.00	60.1	0.0	4.1	0.1	0.2	0.0	0.97	0.00	0.03	0.00	0.00	0.00
HUMA	0.053	0.91	-0.02	0.74	0.28	0.41	-0.23	0.05	0.0	25.4	5.1	22.4	11.0	0.8	0.00	0.61	0.09	0.19	0.06	0.00
SCHE	0.076	0.25	0.32	0.14	0.06	-0.07	-0.09	-0.23	3.9	1.4	0.3	0.8	2.4	21.1	0.42	0.08	0.01	0.02	0.03	0.22
SALA	0.095	0.36	0.30	-0.41	0.17	0.08	-0.01	0.17	4.4	13.7	3.2	1.7	0.0	14.3	0.26	0.47	0.08	0.02	0.00	0.08
SECU	0.088	0.99	0.64	0.02	-0.75	0.06	-0.10	0.04	18.1	0.0	60.6	0.9	3.1	0.7	0.41	0.00	0.57	0.00	0.01	0.00
COMP	0.040	0.58	-0.08	0.49	0.23	-0.22	-0.30	0.04	0.1	8.2	2.5	4.7	14.1	0.3	0.01	0.41	0.09	0.08	0.16	0.00
INTE	0.054	0.48	0.33	0.37	0.00	0.07	0.42	-0.14	3.1	6.6	1.7	0.6	37.2	5.6	0.23	0.29	0.00	0.01	0.37	0.04
NEAR	0.051	0.33	0.06	-0.39	0.16	0.11	-0.06	0.14	0.1	6.8	1.7	1.6	0.8	5.3	0.01	0.46	0.08	0.04	0.01	0.06
ATMO	0.030	0.62	0.16	-0.36	0.39	-0.11	-0.09	-0.41	0.4	3.3	5.4	0.9	1.0	25.8	0.04	0.21	0.24	0.02	0.01	0.28
SOCI	0.037	1.20	0.46	0.60	0.22	-0.71	0.03	0.22	3.9	11.7	2.2	46.4	0.1	8.7	0.17	0.30	0.04	0.41	0.00	0.04
AUTO	0.020	0.48	0.15	0.23	0.03	0.01	0.53	0.05	0.2	0.9	0.0	0.0	22.0	0.3	0.05	0.11	0.00	0.00	0.58	0.01
LIKE	0.057	0.22	-0.01	0.13	0.16	0.15	0.19	0.15	0.0	0.9	1.8	3.4	7.7	6.9	0.00	0.08	0.12	0.11	0.16	0.11
OTHE	0.026	0.34	-0.20	-0.20	-0.27	-0.17	-0.04	-0.16	0.5	0.9	3.3	2.0	0.2	3.3	0.12	0.12	0.22	0.09	0.01	0.07
NONE	0.100	0.37	0.19	-0.48	0.21	-0.09	-0.01	-0.01	1.9	19.9	5.2	2.1	0.0	0.1	0.10	0.62	0.12	0.02	0.00	0.00
OUTD	0.005	2.73	-1.06	-0.18	-0.89	-0.63	-0.03	0.08	3.0	0.2	5.0	5.1	0.0	0.2	0.41	0.01	0.29	0.14	0.00	0.00
NO A	0.016	0.58	0.15	0.08	0.15	0.39	-0.05	-0.14	0.2	0.1	0.4	6.0	0.2	1.7	0.04	0.01	0.04	0.26	0.00	0.04

SUPPLEMENTARY VARIABLES

NAME	MASSES	DIST**2	F1	F2	F3	F4	F5	F6	F1	F2	F3	F4	F5	F6	F1	F2	F3	F4	F5	F6
NODE	0.204	0.23	-0.19	-0.38	0.07	-0.08	-0.04	0.00	0.0	0.0	0.0	0.0	0.0	0.0	0.16	0.62	0.02	0.02	0.01	0.00
GRAD	0.247	0.57	0.24	0.62	0.05	-0.12	0.16	0.04	0.0	0.0	0.0	0.0	0.0	0.0	0.10	0.68	0.00	0.03	0.04	0.00
SMAL	0.342	0.40	-0.52	0.08	0.03	0.07	-0.04	-0.06	0.0	0.0	0.0	0.0	0.0	0.0	0.67	0.02	0.00	0.01	0.00	0.01
MEDI	0.220	0.23	-0.39	-0.01	0.00	-0.07	-0.03	0.04	0.0	0.0	0.0	0.0	0.0	0.0	0.64	0.00	0.00	0.02	0.00	0.01
LARG	0.102	1.40	0.68	-0.39	-0.05	0.06	0.15	0.23	0.0	0.0	0.0	0.0	0.0	0.0	0.33	0.11	0.00	0.00	0.02	0.04

Table 8. Coordinates and contributions of the rows

NAME	MASSES	DIST**2	*	COORDINATES F1	F2	F3	F4	F5	F6	*	ABSOLUTE CONTRIBUTIONS F1	F2	F3	F4	F5	F6	*	SQUARED CORRELATIONS F1	F2	F3	F4	F5	F6	*
FARM	0.085	1.45	*	-1.14	-0.10	-0.32	-0.18	0.07	0.01	*	55.5	0.8	10.8	7.1	1.7	0.0	*	0.89	0.01	0.07	0.02	0.00	0.00	*
FAR2	0.031	0.25	*	0.24	-0.20	0.02	-0.07	-0.13	0.18	*	0.9	1.1	0.0	0.4	2.1	5.2	*	0.24	0.16	0.00	0.02	0.07	0.13	*
ENER	0.015	0.66	*	0.20	-0.15	-0.64	-0.06	0.06	0.25	*	0.3	0.3	7.4	0.1	0.2	4.8	*	0.06	0.03	0.61	0.00	0.01	0.10	*
STEE	0.028	0.62	*	0.39	-0.47	0.33	-0.19	-0.17	-0.03	*	2.1	5.3	3.6	2.5	3.0	0.1	*	0.24	0.35	0.17	0.06	0.04	0.00	*
CHEM	0.017	0.50	*	0.28	-0.34	0.15	0.10	-0.01	0.34	*	0.7	1.7	0.5	0.4	0.0	10.2	*	0.16	0.22	0.04	0.02	0.00	0.23	*
WOOD	0.008	0.86	*	-0.03	0.00	0.34	-0.11	-0.18	-0.29	*	0.0	0.0	1.1	0.2	1.0	3.3	*	0.00	0.00	0.14	0.01	0.04	0.10	*
AUT	0.032	0.54	*	0.57	-0.34	-0.03	-0.02	0.14	0.12	*	5.2	3.2	0.0	0.0	2.3	2.2	*	0.60	0.22	0.00	0.00	0.04	0.02	*
TEXT	0.035	0.39	*	0.12	-0.49	0.26	-0.02	0.04	0.09	*	0.2	7.3	2.8	0.1	0.3	1.4	*	0.03	0.61	0.17	0.01	0.00	0.02	*
PHAR	0.015	0.46	*	0.14	-0.25	0.41	-0.06	-0.07	-0.10	*	0.1	0.8	3.2	0.1	0.3	0.8	*	0.04	0.13	0.37	0.01	0.01	0.02	*
MANU	0.048	0.53	*	0.24	-0.52	0.30	0.08	0.06	0.02	*	1.4	11.3	5.3	0.8	0.7	0.1	*	0.11	0.51	0.17	0.01	0.01	0.00	*
CONS	0.067	0.22	*	-0.19	-0.37	0.08	-0.05	0.07	0.01	*	1.2	8.1	0.5	0.4	1.3	0.0	*	0.16	0.65	0.03	0.01	0.02	0.00	*
FOOD	0.035	0.45	*	-0.44	0.20	0.12	0.31	-0.28	0.06	*	3.4	1.2	0.6	8.4	10.2	0.6	*	0.44	0.09	0.03	0.21	0.17	0.01	*
SBUS	0.072	0.32	*	-0.49	0.12	0.12	0.14	-0.10	-0.02	*	8.9	0.9	1.2	3.4	2.8	0.2	*	0.77	0.05	0.04	0.06	0.03	0.00	*
MBUS	0.040	0.12	*	-0.10	-0.13	0.22	0.10	-0.11	0.02	*	0.2	0.6	2.3	1.0	1.9	0.1	*	0.08	0.14	0.39	0.08	0.10	0.00	*
ADMI	0.080	0.69	*	0.49	0.05	-0.64	0.04	-0.17	-0.00	*	9.5	0.1	40.3	0.3	8.6	8.8	*	0.34	0.00	0.60	0.00	0.04	0.00	*
TELE	0.020	0.54	*	0.42	0.09	-0.40	0.06	-0.01	-0.30	*	1.7	0.1	3.9	0.2	0.0	8.8	*	0.33	0.02	0.30	0.01	0.00	0.16	*
SO.S	0.018	0.73	*	-0.06	0.52	0.31	0.24	-0.32	-0.34	*	0.0	4.1	2.0	2.6	7.1	10.3	*	0.00	0.37	0.13	0.08	0.14	0.16	*
HE.S	0.054	0.35	*	0.11	0.31	-0.02	0.33	0.25	0.06	*	0.3	4.6	0.0	15.1	13.2	0.9	*	0.03	0.28	0.00	0.32	0.18	0.01	*
TEAC	0.081	0.81	*	0.24	0.72	0.20	-0.42	-0.01	0.11	*	2.4	36.2	3.2	35.8	0.1	5.0	*	0.07	0.64	0.05	0.22	0.00	0.02	*
TRAN	0.032	0.33	*	0.28	-0.22	-0.29	-0.10	-0.16	-0.07	*	1.3	1.3	3.9	0.8	3.1	0.7	*	0.24	0.14	0.25	0.03	0.08	0.01	*
BANK	0.034	0.40	*	0.42	0.03	-0.23	0.09	0.18	-0.04	*	3.1	0.0	2.3	0.7	4.4	0.2	*	0.44	0.00	0.13	0.02	0.08	0.00	*
DOME	0.018	0.73	*	0.05	-0.31	0.32	-0.36	-0.26	-0.21	*	0.0	1.5	0.0	5.7	4.5	4.0	*	0.00	0.13	0.14	0.17	0.09	0.06	*
O.SE	0.057	0.24	*	0.09	0.25	0.11	0.03	0.33	-0.05	*	0.2	3.1	0.0	0.8	23.7	0.8	*	0.03	0.27	0.05	0.00	0.46	0.01	*
PRIN	0.027	0.45	*	0.25	-0.02	0.03	-0.09	0.20	-0.53	*	0.8	0.0	0.0	0.5	4.2	38.7	*	0.13	0.00	0.00	0.02	0.09	0.62	*
PRIV	0.034	0.15	*	-0.11	0.13	0.16	0.15	0.06	0.06	*	0.2	0.5	1.1	2.0	0.5	0.6	*	0.08	0.10	0.17	0.15	0.02	0.02	*
NO A	0.020	0.82	*	-0.09	0.59	0.20	0.47	-0.20	0.10	*	0.1	5.9	0.9	11.1	2.9	1.0	*	0.01	0.42	0.05	0.27	0.05	0.01	*

55

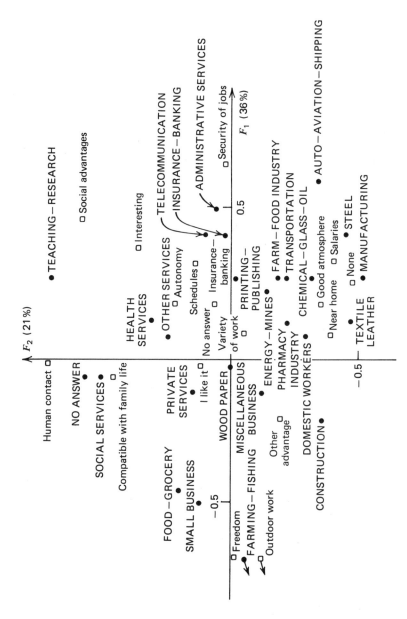

Figure 5. Correspondence analysis: 17 active columns, 26 active rows; projection on axes 1 and 2.

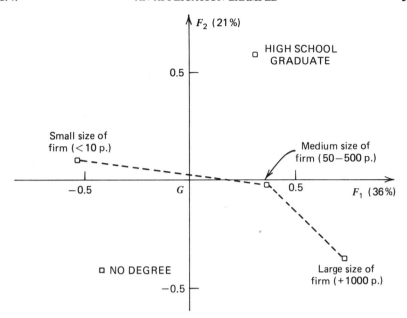

Figure 6.   Positioning of five illustrative columns.

In symmetric fashion, the advantages "human contact" and "compatible with family life" are close because their profiles have percentages that are high for job types "teaching-research," and "social services," and percentages that are low for job types "manufacturing," and "textile-leather."

(d)   Positioning of Supplementary Elements.   To help interpret Figure 5, we can position on the graph the profile-points corresponding to the supplementary columns of Table 5. Figure 6, which can be overlaid on Figure 5, shows the new points. The first axis thus appears to be highly related to the size of the company (cf. the positions of the points "size of firm < 10" on the left, and "size of firm 1000 + " on the right).

The second axis is highly related to the level of education; the point "no degree," bottom left, is opposite from the point "graduate," top right. There is no limit on the number of supplementary points. The positioning is not a costly operation: in the case of this contingency table, it can be done with a pocket calculator. In practice, all available information is positioned on the graph.

Table 9 shows an example of computer printout that allows a quick interpretation of the axes, by showing the placement of the projections of the row-points and column-points among the axes. This table includes both active and supplementary elements. Here we have shown only the positioning on the first axis.

Table 9.   *Aid to interpretation of axes, projection of points on axis 1*

```
         - - - - - - - - - - - - - - - - - - - - - - - - - -    FARM*FARMING/FISHING
OUTD .. OUTDOOR WORK
FREE .. FREEDOM
SMAL .. SMAL/ SIZE OF FIRM   - 10 PERSONS
         - - - - - - - - - - - - - - - - - - - - - - - - - - -    SBUS*SMALL BUSINESS
         - - - - - - - - - - - - - - - - - - - - - - - - - - -    FOOD*FOOD/GROCERY
OTHE .. OTHER
NODE .. NODE/ NO DEGREE
         - - - - - - - - - - - - - - - - - - - - - - - - - -    CONS*CONSTRUCTION
         - - - - - - - - - - - - - - - - - - - - - - - - - -    PRIV*PRIVATE SERVICES
         - - - - - - - - - - - - - - - - - - - - - - - - - -    MBUS*MISCEL.BUSINESS
         - - - - - - - - - - - - - - - - - - - - - - - - -    NO A*NO ANSWER
COMP .. COMPATIBLE WITH FAMILY LIFE
         - - - - - - - - - - - - - - - - - - - - - - - - - -    SO.S*SOCIAL SERVICES
         - - - - - - - - - - - - - - - - - - - - - - - - - -    WOOD*WOOD/PAPER
HUMA .. HUMAN CONTACT
LIKE .. LIKE/ I LIKE IT ,THAT S ALL
         - - - - - - - - - - - - - - - - - - - - - - - - -    DOME*DOMESTIC WORKERS
NEAR .. NEAR HOME
VARI .. VARIETY OF WORK
         - - - - - - - - - - - - - - - - - - - - - - - - -    O.SE*OTHER SERVICES
         - - - - - - - - - - - - - - - - - - - - - - - - -    HE.S*HEALTH SERVICES
         - - - - - - - - - - - - - - - - - - - - - - - - -    TEXT*TEXTILE/LEATHER
         - - - - - - - - - - - - - - - - - - - - - - - - -    PHAR*PHARMACY INDUSTRY
AUTO .. AUTONOMY,I AM MY OWN BOSS
NO A .. NO ANSWER
ATMO .. ATMO/GOOD ATMOSPHERE
NONE .. NONE
         - - - - - - - - - - - - - - - - - - - - - - - - -    ENER*ENERGY/MINES
GRAD .. GRAD/ HIGH SCHOOL GRADUATE
         - - - - - - - - - - - - - - - - - - - - - - - - -    TEAC*TEACHING/RESEARCH
         - - - - - - - - - - - - - - - - - - - - - - - - -    MANU*MANUFACTURING
         - - - - - - - - - - - - - - - - - - - - - - - - -    FAR2*FARM/FOOD INDUSTRY
         - - - - - - - - - - - - - - - - - - - - - - - - -    PRIN*PRINTING/PUBLISHING
         - - - - - - - - - - - - - - - - - - - - - - - - -    CHEM*CHEMICAL/GLASS/OIL
         - - - - - - - - - - - - - - - - - - - - - - - - -    TRAN*TRANSPORTATION
SALA .. SALARIES
SCHE .. SCHEDULES
INTE .. INTERESTING
MEDI .. MEDI/ SIZE OF FIRM 50 - 500 PERSONS
         - - - - - - - - - - - - - - - - - - - - - - - - - - -    STEE*STEEL
         - - - - - - - - - - - - - - - - - - - - - - - - - - -    TELE*TELECOMMUNICATION
         - - - - - - - - - - - - - - - - - - - - - - - - - - -    BANK*INSURANCE/BANKING
SOCI .. SOCIAL ADVANTAGES
         - - - - - - - - - - - - - - - - - - - - - - - - - - -    ADMI*ADMINISTRATIVE SER.
         - - - - - - - - - - - - - - - - - - - - - - - - - - -    AUT *AUTO/AVIATION/SHIP
SECU .. SECURITY OF JOB
LARG .. LARG/ SIZE OF FIRM  + 1000 PERSONS
```

II.5.   COMPUTATIONS

II.5.1.   Analysis with Respect to the Center of Gravity

We perform the computations in $\mathbb{R}^p$.

Each point i, with a weight of $f_{i.}$ has the coordinates

$$\frac{f_{ij}}{f_{i.}} \qquad \text{for } j = 1, \ldots, p$$

The center of gravity G of the set of points, or average point, has, as its jth component,

$$g_j = \sum_{i=1}^{n} f_{i\cdot} \frac{f_{ij}}{f_{i\cdot}} = \sum_{i=1}^{n} f_{ij} = f_{\cdot j} \qquad (65)$$

The analysis with respect to the center of gravity consists of replacing

$$\frac{f_{ij}}{f_{i\cdot}} \quad \text{with} \quad \frac{f_{ij}}{f_{i\cdot}} - f_{\cdot j} \quad \text{or,} \quad \frac{f_{ij} - f_{i\cdot}f_{\cdot j}}{f_{i\cdot}}$$

Note that the set of points is in the $(p - 1)$-dimensional subspace $\mathscr{H}$ defined by the equation

$$\sum_{j} \left(\frac{f_{ij}}{f_{i\cdot}} \right) = 1, \qquad \text{for every } i \qquad (66)$$

This subspace contains the center of gravity G and the principal axes of the analysis with respect to G.

Every vector of this subspace is such that the sum of its components is zero. If $\mathbf{x}$ and $\mathbf{y}$ are two points of $\mathscr{H}$ whose coordinates are x_j and y_j (for $j = 1,\ldots,p$), we have the equation

$$\sum_{j=1}^{p} x_j = \sum_{j=1}^{p} y_j = 1 \quad \text{therefore:} \quad \sum_{j=1}^{p} (x_j - y_j) = 0 \qquad (67)$$

In particular, the components of each of the principal axes add up to zero.

Matrix $\mathbf{S}$ of Section II.2.2. becomes $\mathbf{S}^*$, whose general term is $s_{jj'}^*$, after substituting

$$\frac{f_{ij}}{f_{i\cdot}} \quad \text{with} \quad \frac{f_{ij} - f_{i\cdot}f_{\cdot j}}{f_{i\cdot}}$$

Hence

$$s_{jj'}^* = \sum_{i=1}^{n} \frac{(f_{ij} - f_{i\cdot}f_{\cdot j})(f_{ij'} - f_{i\cdot}f_{\cdot j'})}{f_{i\cdot}f_{\cdot j'}} \qquad (68)$$

By expanding $s_{jj'}^*$, and taking into account the relationships between the three quantities f_{ij}, $f_{i\cdot}$, and $f_{\cdot j}$, we have

$$s_{jj'}^* = \sum_{i} \left(\frac{f_{ij}f_{ij'}}{f_{i\cdot}f_{\cdot j'}} - f_{ij} \right) = s_{jj'} - f_{\cdot j} \qquad (69)$$

If $\mathbf{u}^*$ is a principal axis originating at G, it satisfies

$$\sum_{j'=1}^{p} s_{jj'}^* u_{j'}^* = \lambda^* u_j^* \tag{70}$$

or

$$\sum_{j'} s_{jj'} u_{j'}^* - f_{\cdot j} \sum_{j'} u_{j'}^* = \lambda^* u_j^* \tag{71}$$

Since $\mathbf{u}^* \in \mathcal{H}$, we have $\sum_{j'} u_{j'}^* = 0$, and therefore

$$\sum_{j'} s_{jj'} u_{j'}^* = \lambda^* u_j^* \tag{72}$$

It follows that every eigenvector of $\mathbf{S}^*$ belonging to $\mathcal{H}$ is an eigenvector of $\mathbf{S}$ relative to the same eigenvalue. In addition, the line connecting the origin to G is, as can easily be verified, an eigenvector of $\mathbf{S}$ relative to the eigenvalue 1, and an eigenvector of $\mathbf{S}^*$ relative to the eigenvalue 0.

It is therefore sufficient to diagonalize $\mathbf{S}$ by setting aside the eigenvalue equal to 1 and its corresponding axes in $\mathbb{R}^p$ and $\mathbb{R}^n$.

The *trace* of $\mathbf{S}^*$ has a remarkable property. According to equation (68) we have

$$\operatorname{tr} \mathbf{S}^* = \sum_{i=1}^{n} \frac{\left(f_{ij} - f_{i\cdot} f_{\cdot j}\right)^2}{f_{i\cdot} f_{\cdot j}} \tag{73}$$

Since the total is k, we see that the quantity

$$k \operatorname{tr} \mathbf{S}^*$$

is the statistic that follows an approximate chi-square distribution with $(n-1)(p-1)$ degrees of freedom, testing the hypothesis of independence of the rows and columns of matrix $\mathbf{K}$. Thus the sum of the nontrivial eigenvalues of a correspondence analysis has a statistical interpretation. For matrix $\mathbf{S}$, we have the relationship $\operatorname{tr} \mathbf{S} = \operatorname{tr} \mathbf{S}^* + 1$. On the other hand, in normalized principal components analysis (Chapter I), the sum of the eigenvalues is constant (the trace of the correlation matrix).

II.5.2. Symmetrization of S

Matrix $\mathbf{S}$ is not symmetric. Its general term is written

$$s_{jj'} = \sum_{i=1}^{n} \frac{f_{ij} f_{ij'}}{f_{i\cdot} f_{\cdot j'}} \tag{74}$$

In matrix notation (cf. Section II.2.2),

$$S = F'D_n^{-1}FD_p^{-1}. \tag{75}$$

Matrix $\hat{A} = F'D_n^{-1}F$ is symmetric and matrix D_p^{-1} is diagonal. We may write

$$\frac{1}{f_{\cdot j}} = \left(\frac{1}{\sqrt{f_{\cdot j}}}\right)\left(\frac{1}{\sqrt{f_{\cdot j}}}\right) \tag{76}$$

or

$$D_p^{-1} = D_p^{-1/2}D_p^{-1/2} \tag{77}$$

Hence

$$S = \hat{A}D_p^{-1/2}D_p^{-1/2} \tag{78}$$

Starting with equation $Su = \lambda u$, or

$$\hat{A}D_p^{-1/2}D_p^{-1/2}u = \lambda u \tag{79}$$

let us premultiply the two members by $D_p^{-1/2}$, and set $D_p^{-1/2}u = w$. We obtain

$$D_p^{-1/2}\hat{A}D_p^{-1/2}w = \lambda w \tag{80}$$

The matrix

$$S_1 = D_p^{-1/2}\hat{A}D_p^{-1/2} \tag{81}$$

is symmetric, and has the same eigenvalue λ as S. It is convenient to diagonalize it, then to calculate u from the equation

$$u = D_p^{1/2}w \tag{82}$$

Factor φ is obtained from

$$\varphi = D_p^{-1}u = D_p^{-1/2}w \tag{83}$$

Note. We could have performed the transformation

$$y = D_p^{-1/2} x \tag{84}$$

on the initial coordinates in $\mathbb{R}^p$.

The chi-square distance between two points x_1 and x_2,

$$d^2(x_1, x_2) = (x_1 - x_2)' D_p^{-1} (x_1 - x_2) \tag{85}$$

becomes, with the new coordinates, the usual distance,

$$d^2(y_1, y_2) = (y_1 - y_2)'(y_1 - y_2) \tag{86}$$

This leads directly to the diagonalization of the symmetric matrix S_1.

Canonical Analysis and Discriminant Analysis— Theoretical and Technical Considerations: A Brief Review

In this chapter we discuss two different statistical techniques, emphasizing their connection with correspondence analysis.

1. *Canonical analysis* has an important theoretical role; it bridges the gap between multiple regression, discriminant analysis, and correspondence analysis.

2. *Multiple discriminant analysis* has many more applications than canonical analysis, particularly in medicine and biometry. However, it is being supplanted by other discriminant methods that rely on less rigid assumptions.

The mathematical proofs of these methods are not generally essential for the reader whose main interest lies in data description; however they are quite helpful in familiarizing the reader with the algebraic and geometric bases of linear data analysis.

III.1. CANONICAL ANALYSIS

Canonical analysis (or *canonical correlation analysis*) was developed by Hotelling (1936). Its applications are rather limited, since interpretation of the analysis' results is not generally straightforward. However, it provides a general theoretical basis for multiple regression and discriminant analysis, which are special cases of canonical analysis. *Canonical analysis* attempts to identify relationships between *two sets* of variables, by finding the linear combinations of the variables in the first set that are most highly correlated with the linear combinations of the variables in the second set. (This reduces to multiple regression if one of the two sets is a single variable.)

III.1.1. Notation and Formulation of the Problem

The data matrix R, with n rows and $p + q$ columns is partitioned into two submatrices X and Z, that have p and q columns, respectively.

$$R = [X|Z]$$

The rows represent individual respondents or observations; the first p columns are the variables in the first set and the following q columns are the variables in the second set.

Let us assume that the variables are *centered*, in other words, the sum of the elements of each column of R is equal to 0. Then the matrix of sample covariances of the $p + q$ variables is written

$$V(R) = \frac{1}{n} R'R \qquad \left(v_{jj'} = \frac{1}{n} \sum_k r_{kj} r_{kj'} \right) \tag{1}$$

Namely,

$$V(R) = \frac{1}{n} \begin{bmatrix} X'X & X'Z \\ Z'X & Z'Z \end{bmatrix} \tag{2}$$

Let us consider observation i, which is the ith row of R:

$$\left(x_{i1}, x_{i2}, \ldots, x_{ip}, z_{i1}, z_{i2}, \ldots, z_{iq} \right) \tag{3}$$

Let a and b be two vectors with p and q components that define two linear combinations $a(i)$ and $b(i)$:

$$a(i) = \sum_{j=1}^{p} a_j x_{ij} \tag{4}$$

$$b(i) = \sum_{j=1}^{q} b_j z_{ij} \tag{5}$$

The n values of $a(i)$ for all i observations are components of $\mathbf{Xa}$. Similarly, the n values of $b(i)$ are the components of $\mathbf{Zb}$.

We search for the two linear combinations $a(i)$ that are the most highly correlated over all the values of i. These two linear combinations are called *canonical variables*.

Since the initial variables are centered, their linear combinations are also centered.

Given that correlation coefficients are independent of the scale of the variables, we impose the constraint of *unit variance* on the two linear combinations.

The variance of all of the values of $a(i)$ for $i = 1, 2, \ldots, n$ is writen var($\mathbf{a}$):

$$\text{var}(\mathbf{a}) = \frac{1}{n} \sum_{i=1}^{n} a^2(i) = \frac{1}{n}(\mathbf{Xa})'\mathbf{Xa} = \frac{1}{n}\mathbf{a'X'Xa} \tag{6}$$

Similarly,

$$\text{var}(\mathbf{b}) = \frac{1}{n}\mathbf{b'Z'Zb} \tag{7}$$

Under these circumstances the correlation coefficient between the linear combinations $a(i)$ and $b(i)$ is simply the covariance

$$\text{cov}(\mathbf{a}, \mathbf{b}) = \frac{1}{n} \sum_{i=1}^{n} a(i)b(i) \tag{8}$$

Namely,

$$\text{cov}(\mathbf{a}, \mathbf{b}) = \frac{1}{n}\mathbf{a'X'Zb} \tag{9}$$

Finally, the mathematical problem of finding the maximum correlation is written, after dropping $1/n$ from the equations (remember that $\mathbf{X}$ and $\mathbf{Z}$ are centered): find $\mathbf{a}$ and $\mathbf{b}$ that maximize

$$\mathbf{a'X'Zb}$$

with the constraints

$$\begin{cases} \mathbf{a'X'Xa} = 1 \\ \mathbf{b'Z'Zb} = 1 \end{cases}$$

III.1.2.   Calculation of the Canonical Variables

This derivation is analogous to the one developed in Section I.2. Two Lagrangian multipliers are used; the quantity to maximize is

$$\mathcal{L} = \mathbf{a'X'Zb} - \lambda(\mathbf{a'X'Xa} - 1) - \mu(\mathbf{b'Z'Zb} - 1) \tag{10}$$

Setting the partial derivatives equal to zero yields the following system of equations:

$$\mathbf{X'Zb} - 2\lambda\mathbf{X'Xa} = 0$$

$$\mathbf{Z'Xa} - 2\mu\mathbf{Z'Zb} = 0 \tag{11}$$

We premultiply the elements of these two equations by $\mathbf{a'}$ and $\mathbf{b'}$, respectively. Taking into account the constraints

$$\mathbf{a'X'Xa} = \mathbf{b'Z'Zb} = 1 \tag{12}$$

these equations simplify into

$$\mathbf{a'X'Zb} = 2\lambda$$

$$\mathbf{b'Z'Xa} = 2\mu \tag{13}$$

Consequently, $\lambda = \mu$. Let us set

$$\beta = 2\lambda \tag{14}$$

Note that β is the *maximum correlation coefficient* that is required. The system of equations is then written

$$\mathbf{X'Zb} = \beta\mathbf{X'Xa} \tag{15}$$

$$\mathbf{Z'Xa} = \beta\mathbf{Z'Zb} \tag{16}$$

To solve this system of equations, the two matrices, $\mathbf{X'X}$ and $\mathbf{Z'Z}$, must be *nonsingular*. By extracting the value for $\mathbf{a}$ from equation (15) and substituting in equation (16), we obtain

$$\mathbf{Z'X(X'X)^{-1}X'Zb} = \beta^2\mathbf{Z'Zb} \tag{17}$$

This shows that $\mathbf{b}$ is an eigenvector of

$$\mathbf{(Z'Z)^{-1}Z'X(X'X)^{-1}X'Z} \tag{18}$$

relative to the greatest eigenvalue β^2, which is the square of the correlation coefficient between the linear combinations $\mathbf{a}$ and $\mathbf{b}$.

This value, β^2, is the first *canonical root* or the square of the first *canonical correlation coefficient* between the two variables.

a can be calculated either from equation (15) and **b**, or directly from the following equation, which is obtained in the same way as equation (17):

$$\mathbf{X'Z(Z'Z)}^{-1}\mathbf{Z'Xa} = \beta^2 \mathbf{X'Xa} \tag{19}$$

We can generalize this result as follows: the eigenvectors (in order of decreasing size of the corresponding eigenvalues) correspond to the pairs of linear combinations that are the most highly correlated among themselves, given that within each group of variables, successive linear combinations must be uncorrelated.

(a) Relationship with Multiple Regression. Suppose matrix **Z** has only one column ($q = 1$). Then **b** has just one component, also called b. In this circumstance **Z'Z** is a scalar and equation (17) becomes

$$\beta^2 = \frac{\mathbf{Z'X(X'X)}^{-1}\mathbf{X'Z}}{\mathbf{Z'Z}} \tag{20}$$

β^2 is the multiple correlation coefficient between column matrix **Z** (endogenous or explained variable) and the columns of matrix **X** (exogenous or explanatory variables).

In this case equation (15) is written

$$\mathbf{a} = \frac{b}{\beta}(\mathbf{X'X})^{-1}\mathbf{X'Z} \tag{21}$$

This equation shows that vector **a** is proportional (coefficient b/β) to the vector of the coefficients of the multiple regression that explains variable **Z** through the p variables (columns of **X**).

In fact, the coefficient b/β is easily calculated, since

$$b = \frac{1}{\sqrt{\mathbf{Z'Z}}} \tag{22}$$

because of the constraint of normalization.

(b) Geometric Interpretation. Equations (15) and (16) may be written

$$\mathbf{a} = \frac{1}{\beta}(\mathbf{X'X})^{-1}\mathbf{X'Zb} \tag{23}$$

$$\mathbf{b} = \frac{1}{\beta}(\mathbf{Z'Z})^{-1}\mathbf{Z'Xa} \tag{24}$$

By premultiplying the two elements of each equation by $\mathbf{X}$ and $\mathbf{Z}$, respectively, we obtain

$$\mathbf{Xa} = \frac{1}{\beta}\mathbf{X}(\mathbf{X}'\mathbf{X})^{-1}\mathbf{X}'\mathbf{Zb} \tag{25}$$

$$\mathbf{Zb} = \frac{1}{\beta}\mathbf{Z}(\mathbf{Z}'\mathbf{Z})^{-1}\mathbf{Z}'\mathbf{Xa} \tag{26}$$

Let us call $\mathscr{V}_X$ and $\mathscr{V}_Z$ the linear subspaces of $\mathbb{R}^n$ that are generated by the columns of $\mathbf{X}$ and $\mathbf{Z}$, respectively.

The linear combinations $\mathbf{a}$ and $\mathbf{b}$ define the points of $\mathscr{V}_X$ and of $\mathscr{V}_Z$ whose coordinates are the components of the vectors $\mathbf{Xa}$ and $\mathbf{Zb}$, respectively.

The symmetric and idempotent matrices

$$\mathbf{P}_X = \mathbf{X}(\mathbf{X}'\mathbf{X})^{-1}\mathbf{X}' \tag{27}$$

$$\mathbf{P}_Z = \mathbf{Z}(\mathbf{Z}'\mathbf{Z})^{-1}\mathbf{Z}' \tag{28}$$

are the projection operators on $\mathscr{V}_X$ and $\mathscr{V}_Z$, respectively. In other words, equations (25) and (26) express that each vector is collinear to the projection of the other. See Figure 1.

Since vectors $\mathbf{Xa}$ and $\mathbf{Xb}$ are unit vectors, the formulas show that

$$\beta = \cos \omega = \cos(\mathbf{Xa}, \mathbf{Zb}) \tag{29}$$

It becomes evident that the first canonical root β^2 is the squared cosine of the *smallest angle* between the subspaces $\mathscr{V}_X$ and $\mathscr{V}_Z$ (cf. Figure 1).

With these geometric considerations in mind, we can write equations (25) and (26) directly and thus compute the canonical variables directly. For example, in equation (26) $\mathbf{Xa}$ is substituted for its value taken from equation (25).

(c)   Case of Singular Matrices.   Let us examine the situation where $\mathbf{X}'\mathbf{X}$ and $\mathbf{Z}'\mathbf{Z}$ are singular. Take $\mathbf{Z}'\mathbf{Z}$, for example. Singularity means that the (n, q) matrix $\mathbf{Z}$ has a rank smaller than q; let $q - s$ be its rank.

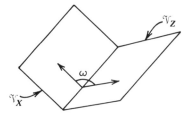

Figure 1

As in regression, there are two ways of proceeding to solve the system of matrix equations (15) and (16):

1.   In $\mathbb{R}^n$, we choose a basis of the subspace $\mathscr{V}_Z$ with $q - s$ dimensions, generated by $\mathbf{Z}$. This basis is described by the $q - s$ columns of a matrix $\mathbf{Z}^*$ (it is better to choose an orthogonal basis, obtained by the Gram-Schmidt procedure, or by general analysis of $\mathbf{Z}$); and in the computations $\mathbf{Z}\mathbf{b}$ is substituted with $\mathbf{Z}^*\mathbf{b}^*$ where $\mathbf{b}^*$ is a vector with $q - s$ components. Now matrix $\mathbf{Z}^{*\prime}\mathbf{Z}^*$ can be inverted.

2.   As is frequently done in the case of the general linear model, a *full rank* matrix $\mathbf{Z}_0$ is built, such that $\mathscr{V}_Z \subset \mathscr{V}_{Z_0}$. In order to recover the original linear subspace $\mathscr{V}_Z$, we must then impose a constraint on $\mathbf{b}$, namely: $\mathbf{Z}_0\mathbf{b}$ *must belong* to $\mathscr{V}_Z$. If $\mathbf{Z}_1$ designates an (n, s) matrix, such that $\mathbf{Z}_1'\mathbf{Z} = 0$ and such that $\mathbf{Z}_1\mathbf{b} \in \mathscr{V}_{Z_0}$, the constraint on $\mathbf{b}$ is written

$$\mathbf{Z}_1'\mathbf{Z}_0\mathbf{b} = 0 \tag{30}$$

Note.   This situation is going to show up in a particularly simple context in *discriminant analysis* (cf. Section III.2.3): the (n, q) matrix $\mathbf{Z}$ is singular, whereas the initial matrix $\mathbf{Z}_0$ (prior to centering) is of full rank. This results from the fact that the mapping generated by $\mathbf{Z}_0$ contains the vector $\mathbf{e}_n$ of $\mathbb{R}^n$, all of whose components are equal to 1. We can then deal with $\mathbf{Z}_0$, knowing that $\mathbf{b}$ is subject to

$$\mathbf{e}_n'\mathbf{Z}_0\mathbf{b} = 0 \tag{31}$$

which is written

$$\sum_{j=1}^{q} z_{.j}b_j = 0 \tag{32}$$

where $z_{.j}$ designates the sum of column j of matrix $\mathbf{Z}_0$.

III.2.   DISCRIMINANT ANALYSIS

Discriminant analysis is a group of techniques whose purpose is to describe and classify individuals on whom a large number of variables have been measured. We limit ourselves to outlining the principal method, and to showing its relationship with canonical analysis (of which it is a special case) and correspondence analysis of contingency tables (which is one of its special cases, where only dummy variables are involved).

Early work on this topic dates back to Fisher (1936), or, more indirectly, to Mahalanobis (1936). Another formulation can be found in Rao (1952).

III.2.1. Formulation of Problem

Multiple discriminant analysis addresses the problem of assigning individuals (characterized by description vectors) to certain groups that are already identified in the sample. This assignment problem is different from the problem of classification considered in Chapter V, which consists of building the most homogeneous possible groups in a sample.

Let us consider, for example, a data matrix that contains the annual expenditures of 1000 households allocated into 100 variables. Let us assume that we also have a partitioning of the 1000 households according to 9 socioeconomic categories of the head of household.

We can ask the following question: given an additional household for which the 100 types of annual consumption are known, is it possible to *predict* its socioeconomic category?

The question we have formulated here is artificial; it does not have a practical purpose. However it may happen that measurements on a number of variables for an individual are the only means of assigning him or her to a particular group. This is sometimes true in medicine where numerous examinations or analyses are the only means of assessing whether a patient has a certain disease, and whether an operation is required. The techniques of discriminant analysis can make a diagnosis possible, and provide information that is useful in choosing alternatives.

Multiple discriminant analysis finds a set of *linear combinations of the variables* whose values are as close as possible *within* groups and as far apart as possible *between* groups.

We elaborate on these ideas and introduce a few definitions and some notation.

Let $Y = [y_{ij}]$ be the data matrix with n rows (individuals or observations) and p columns (variables).

The overall mean $\bar{y}_j$ of variable j is written

$$\bar{y}_j = \frac{1}{n} \sum_{i=1}^{n} y_{ij} \tag{33}$$

The n rows of Y are partitioned a priori into q groups. Group k is characterized by a set I_k of n_k values of index i, with

$$\sum_{k=1}^{q} n_k = n \tag{34}$$

Let $\bar{y}_{kj}$ be the mean of variable j in group k:

$$\bar{y}_{kj} = \frac{1}{n_k} \sum_{i \in I_k} y_{ij} \tag{35}$$

For every variable j we have the equation

$$\bar{y}_j = \sum_{k=1}^{q} \frac{n_k}{n} \bar{y}_{kj} \tag{36}$$

The total covariance between two variables j and j' is written

$$\text{cov}(j, j') = \frac{1}{n} \sum_{i=1}^{n} (y_{ij} - \bar{y}_j)(y_{ij'} - \bar{y}_{j'}) \tag{37}$$

or

$$\text{cov}(j, j') = \frac{1}{n} \sum_{k=1}^{q} \left[\sum_{i \in I_k} (y_{ij} - \bar{y}_j)(y_{ij'} - \bar{y}_{j'}) \right] \tag{38}$$

As in analysis of variance, we partition $\text{cov}(j, j')$ into the sum of the within-group covariances and between-group covariances.

To do this we begin with the following identity for $i, j,$ and k:

$$(y_{ij} - \bar{y}_j) = (y_{ij} - \bar{y}_{kj}) + (\bar{y}_{kj} - \bar{y}_j) \tag{39}$$

The bracketed sum in the covariance formula then factorizes into four terms, two of which are zero. This is because, by the definition of $\bar{y}_{kj}$;

$$\sum_{i \in I_k} (y_{ij} - \bar{y}_{kj})(\bar{y}_{kj'} - \bar{y}_{j'}) = (\bar{y}_{kj'} - \bar{y}_{j'}) \sum_{i \in I_k} (y_{ij} - \bar{y}_{kj}) = 0 \tag{40}$$

Similarly;

$$\sum_{i \in I_k} (\bar{y}_{kj} - \bar{y}_j)(y_{ij'} - \bar{y}_{kj'}) = 0 \tag{41}$$

This leaves us with the so-called Huyghens decomposition formula, or the analysis of variance equation:

$$\text{cov}(j, j') = \frac{1}{n} \sum_{k=1}^{q} \sum_{i \in I_k} (y_{ij} - \bar{y}_{kj})(y_{ij'} - \bar{y}_{kj'}) +$$

$$\sum_{k=1}^{q} \frac{n_k}{n} (\bar{y}_{kj} - \bar{y}_j)(\bar{y}_{kj'} - \bar{y}_{j'}) \tag{42}$$

We write this equation in matrix form (the *total* covariance equals the sum of the covariance *within* the groups and the covariance *between* the groups):

$$T = W + B \tag{43}$$

with

$$t_{jj'} = \text{cov}(j, j') \tag{44}$$

$$w_{jj'} = \frac{1}{n} \sum_{k=1}^{q} \sum_{i \in I_k} (y_{ij} - \bar{y}_{kj})(y_{ij'} - \bar{y}_{kj'}) \tag{45}$$

$$b_{jj'} = \sum_{k=1}^{q} \frac{n_k}{n} (\bar{y}_{kj} - \bar{y}_j)(\bar{y}_{kj'} - \bar{y}_{j'}) \tag{46}$$

Now, let $a(i)$ be the value of a linear combination of the p centered variables for individual i:

$$a(i) = \sum_{j=1}^{p} a_j(y_{ij} - \bar{y}_j) \tag{47}$$

The variance var(**a**) of the derived variable $a(i)$ has the following value, since $a(i)$ is centered:

$$\text{var}(\mathbf{a}) = \frac{1}{n} \sum_{i=1}^{n} a^2(i) = \frac{1}{n} \sum_{i=1}^{n} \left[\sum_{j=1}^{p} a_j(y_{ij} - \bar{y}_j) \right]^2 \tag{48}$$

$$\text{var}(\mathbf{a}) = \frac{1}{n} \sum_{i=1}^{n} \sum_{j=1}^{p} \sum_{j'=1}^{p} a_j a_{j'}(y_{ij} - \bar{y}_j)(y_{ij'} - \bar{y}_{j'}) \tag{29}$$

By changing the order of the summations, we obtain

$$\text{var}(\mathbf{a}) = \sum_{j=1}^{p} \sum_{j'=1}^{p} a_j a_{j'} \text{cov}(j, j') = \mathbf{a}'\mathbf{T}\mathbf{a} \tag{50}$$

where **a** is the vector whose p components are $a_1, \ldots, a_p$.

Thus the variance of a linear combination **a** of the variables is partitioned, according to equation (43), into within variance and between variance:

$$\mathbf{a}'\mathbf{T}\mathbf{a} = \mathbf{a}'\mathbf{W}\mathbf{a} + \mathbf{a}'\mathbf{B}\mathbf{a} \tag{51}$$

The problem addressed by multiple discriminant analysis can then be formulated as follows.

Among all of the linear combinations of the variables let us find those that have a maximum between groups variance (in order to maximize the differences between groups) and a minimum within groups variance (in order to minimize the spread of the groups). These linear combinations are the linear discriminant functions.

We must find $\mathbf{a}$ such that the quotient $\mathbf{a'Ba}/\mathbf{a'Wa}$ is maximum (or $\mathbf{a'Wa}/\mathbf{a'Ba}$ minimum). According to equation (51), minimizing $\mathbf{a'Ta}/\mathbf{a'Ba}$ is equivalent to maximizing

$$f(\mathbf{a}) = \frac{\mathbf{a'Ba}}{\mathbf{a'Ta}} \tag{52}$$

III.2.2.   Calculation of the Linear Discriminant Functions

Given that the function $f(\mathbf{a})$ is homogeneous and zeroth degree in $\mathbf{a}$ (that is, invariant if $\mathbf{a}$ is changed to $\mu\mathbf{a}$, where μ is a scalar), the problem is the same as finding the maximum of the quadratic form $\mathbf{a'Ba}$ with the quadratic constraint $\mathbf{a'Ta} = 1$.

We have the following matrix equation (cf. Section I.5.1):

$$2\mathbf{Ba} - 2\lambda\mathbf{Ta} = 0 \tag{53}$$

or

$$\mathbf{Ba} = \lambda\mathbf{Ta} \tag{54}$$

Generally the covariance matrix $\mathbf{T}$ is nonsingular. Therefore,

$$\mathbf{T}^{-1}\mathbf{Ba} = \lambda\mathbf{a} \tag{55}$$

$\mathbf{a}$ is the eigenvector of $\mathbf{T}^{-1}\mathbf{B}$ relative to the largest eigenvalue λ. When we premultiply the two elements of equation (54) by $\mathbf{a'}$, we observe that $\mathbf{a'Ba}$, the required maximum, is none other than λ.

The largest eigenvalue λ, which is the quotient of the *between-groups* variance of the discriminant function and the *total* variance, is less than 1 according to equation (51). It is sometimes called the *discriminating power* of function $\mathbf{a}$.

Diagonalizing a Symmetric Matrix. Matrix $\mathbf{T}^{-1}\mathbf{B}$ is not symmetric. However, it is possible to transform the problem into diagonalizing a symmetric (q, q) matrix (recall that p is the number of variables and q is the number of groups, generally much smaller than p).

Matrix **B**, such that

$$b_{jj'} = \sum_{k=1}^{q} \frac{n_k}{n} (\bar{y}_{kj} - \bar{y}_j)(\bar{y}_{kj'} - \bar{y}_{j'}) \tag{56}$$

may be considered the product of a matrix $\mathbf{B}_1$ with p rows and q columns by its transpose; $\mathbf{B}_1 = [b_1(j, k)]$ is such that

$$b_1(j, k) = \sqrt{\frac{n_k}{n}} (\bar{y}_{kj} - \bar{y}_j) \tag{57}$$

Equation (54) is written

$$\mathbf{B}_1 \mathbf{B}_1' \mathbf{a} = \lambda \mathbf{T} \mathbf{a} \tag{58}$$

Suppose

$$\mathbf{a} = \mathbf{T}^{-1} \mathbf{B}_1 \mathbf{v} \tag{59}$$

Equation (54) is then written

$$\mathbf{B}_1 \mathbf{B}_1' \mathbf{T}^{-1} \mathbf{B}_1 \mathbf{v} = \lambda \mathbf{B}_1 \mathbf{v} \tag{60}$$

Every eigenvector **v** relative to an eigenvalue λ (different from 0) of the symmetric (q, q) matrix;

$$\mathbf{B}_1' \mathbf{T}^{-1} \mathbf{B}_1 \tag{61}$$

clearly satisfes equation (60) as well. Vector **a** and scalar λ then satisfy equation (54).

In practice, this symmetric matrix is first diagonalized, and then **a** is derived from it by the transformation $\mathbf{a} = \mathbf{T}^{-1}\mathbf{B}_1\mathbf{v}$. The symmetric (q, q) matrix is generally considerably smaller than the nonsymmetric matrix $\mathbf{T}^{-1}\mathbf{B}$.

III.2.3. Relationship with Canonical Analysis

Let us continue to designate the data matrix with n rows and p columns as matrix **Y**; and let us call $\mathbf{X} = [x_{ij}]$ the matrix of the centered variables such that

$$x_{ij} = y_{ij} - \bar{y}_j \tag{62}$$

We code the information relative to the partition of the n individuals into q groups by building a matrix $\mathbf{Z}$ with n rows and q columns: element z_{ik} of the kth column of $\mathbf{Z}$ takes the value 1 if individual i belongs to group k and 0 otherwise.

As before, let us suppose

$$\mathbf{R} = [\mathbf{X}|\mathbf{Z}] \tag{63}$$

In other words, as in analysis of covariance, we add the initial variables new, dummy variables that are indicators of membership to the various groups.

Note that contrary to Section III.1, the columns of $\mathbf{Z}$ are not centered: the sum of the elements of the kth column is n_k.

The blocks of the matrix are as follows:

$$\mathbf{V}(\mathbf{R}) = \frac{1}{n}\mathbf{R'R} = \frac{1}{n}\begin{bmatrix} \mathbf{X'X} & \mathbf{X'Z} \\ \mathbf{Z'X} & \mathbf{Z'Z} \end{bmatrix} \tag{64}$$

We take into account the special nature of the columns of $\mathbf{Z}$.

Matrix $(1/n)\mathbf{X'X}$ is the covariance matrix we previously called $\mathbf{T}$.

Matrix $\mathbf{D} = \mathbf{Z'Z}$ is diagonal, and its kth element is n_k, which is the size of the kth group. This is because

$$\sum_{i=1}^{n} z_{ik}z_{ik'} = \delta_{kk'}n_k \tag{65}$$

since individual i belongs either to group k or group k'. For $k = k'$, there are as many nonzero terms as there are individuals in group k.

Matrix $\mathbf{G} = \mathbf{X'Z}$, with p rows and q columns, is such that

$$g_{jk} = \sum_{i=1}^{n} x_{ij}z_{ik} = \sum_{i=1}^{n} (y_{ij} - \bar{y}_j)z_{ik} = \sum_{i \in I_k} (y_{ij} - \bar{y}_j) \tag{66}$$

or

$$g_{jk} = n_k(\bar{y}_{kj} - \bar{y}_j) \tag{67}$$

Matrix $\mathbf{B}$ defined in equation (43) is therefore expressed as

$$\mathbf{B} = \frac{1}{n}\mathbf{GD}^{-1}\mathbf{G'} \tag{68}$$

or

$$B = \frac{1}{n}X'Z(Z'Z)^{-1}Z'X \qquad (69)$$

The fundamental relationship of discriminant analysis, equation (54), is now written

$$X'Z'(Z'Z)^{-1}Z'Xa = \lambda X'Xa \qquad (70)$$

This is the same as equation (19) of Section III.1.2 if we set $\lambda = \beta^2$. It is evident that the linear discriminant function **a** can be obtained by a canonical analysis of matrix $R = [X|Z]$ where Z is now the noncentered matrix. (Z was called Z_0 at the end of Section III.1.2.)

Furthermore, for any value of i, we have

$$\sum_{k=1}^{q} z_{ik} = 1 \qquad (71)$$

The subspace $\mathcal{V}_Z$ then contains the vector of $\mathbb{R}^n$ all of whose components equal 1 and that is designated by e_n.

If we designate by e_q the vector whose q components equal 1, we have

$$Ze_q = e_n \qquad (72)$$

After centering, matrix Z becomes

$$\hat{Z} = \left(I - \frac{1}{n}e_ne_n'\right)Z \qquad (73)$$

It then satisfies the equation

$$\hat{Z}e_q = 0 \qquad (74)$$

Note that $(I - (1/n)e_ne_n')$ is a projection matrix that was called P in Section I.3.2.

Matrix $\hat{Z}'\hat{Z}$ is therefore *singular*. Since $\mathcal{V}_{\hat{Z}} \subset \mathcal{V}_Z$, the same result is obtained by analyzing Z, which is not centered; and by keeping among the vectors Zb those that satisfy

$$e_n'Zb = 0 \qquad (75)$$

that is,

$$\sum_{k=1}^{q} z_{.k}b_k = 0 \qquad \text{with } z_{.k} = \sum_{i=1}^{n} z_{ik} \qquad (76)$$

Actually, only vector **a** is of interest during this canonical analysis.

We also note that this analysis requires the same normalization constraint: $\mathbf{a}'\mathbf{T}\mathbf{a} = 1$.

Discriminant analysis is therefore a special case of canonical analysis when one of the two sets of variables is comprised of Boolean vectors (vectors consisting of 0's or 1's), each of which describes a group of the partition of the sample.

III.2.4. Case of Two Groups

This case, a frequent one in practice, leads to many simplifications.

Let us give the two groups the indices 1 and 2. Then the matrix of between group covariances is such that

$$b_{jj'} = \frac{n_1}{n}(\bar{y}_{1j} - \bar{y}_j)(\bar{y}_{1j'} - \bar{y}_{j'}) + \frac{n_2}{n}(\bar{y}_{2j} - \bar{y}_j)(\bar{y}_{2j'} - \bar{y}_{j'}) \qquad (77)$$

with

$$\bar{y}_j = \frac{n_1}{n}\bar{y}_{1j} + \frac{n_2}{n}\bar{y}_{2j} \qquad (78)$$

Substituting $\bar{y}_j$ with its value, and taking into account that $n_1 + n_2 = n$, we find

$$b_{jj'} = \frac{n_1 n_2}{n^2}(\bar{y}_{1j} - \bar{y}_{2j})(\bar{y}_{1j'} - \bar{y}_{2j'}) \qquad (79)$$

The symmetric (p, p) matrix $\mathbf{B}$ may be considered the product of a column matrix $\mathbf{c}$ by its transpose:

$$\mathbf{B} = \mathbf{c}\mathbf{c}' \qquad (80)$$

with

$$c_j = \frac{\sqrt{n_1 n_2}}{n}(\bar{y}_{1j} - \bar{y}_{2j}) \qquad (81)$$

Equation (55) is written

$$\mathbf{T}^{-1}\mathbf{c}\mathbf{c}'\mathbf{a} = \lambda\mathbf{a} \qquad (82)$$

Let us premultiply the two members by $\mathbf{c}'$:

$$[\mathbf{c}'\mathbf{T}^{-1}\mathbf{c}]\mathbf{c}'\mathbf{a} = \lambda\mathbf{c}'\mathbf{a} \qquad (83)$$

The bracketed quantity is a scalar, which is therefore equal to λ which is the only eigenvalue here.

This eigenvalue;

$$\lambda = \mathbf{c}'\mathbf{T}^{-1}\mathbf{c} \tag{84}$$

is closely related to the *generalized distance* between two groups, or the "Mahalanobis D^2." The corresponding eigenvector;

$$\mathbf{a} = \mathbf{T}^{-1}\mathbf{c} \tag{85}$$

is the only discriminant function.

Discriminant analysis is then a special case of multiple regression where the dependent variable has only two values, one for each group.

Let us consider a vector $\mathbf{v}$ with n components such that

$$v_i = \begin{cases} \sqrt{\dfrac{n_1}{n_2}} & \text{if the } i\text{th individual belongs to group 1} \\[2ex] -\sqrt{\dfrac{n_2}{n_1}} & \text{if the } i\text{th individual belongs to group 2} \end{cases}$$

The multiple regression that explains $\mathbf{v}$ with the columns of $\mathbf{Y}$ yields the following coefficients:

$$\mathbf{a} = (\mathbf{Y}'\mathbf{Y})^{-1}\mathbf{Y}'\mathbf{v} \tag{86}$$

.with

$$\frac{1}{n}\mathbf{Y}'\mathbf{Y} = \mathbf{T} \tag{87}$$

and, as can readily be shown,

$$\frac{1}{n}\mathbf{Y}'\mathbf{v} = \mathbf{c} \tag{88}$$

Hence

$$\mathbf{a} = \mathbf{T}^{-1}\mathbf{c} \tag{89}$$

The vector of regression coefficients $\mathbf{a}$ thus coincides with the vector of the components of the discriminant function that was previously computed.

III.2.5. Relationship with Correspondence Analysis

We have seen that multiple discriminant analysis is a special case of canonical analysis of a matrix $\mathbf{R} = [\mathbf{X}|\mathbf{Z}]$ when the submatrix $\mathbf{Z}$ has columns that are the q indicator variables of a partition. In addition we have seen that in terms of the required linear combinations it is equivalent (and simpler analytically) not to center the columns of $\mathbf{Z}$.

When matrix $\mathbf{X}$ also describes a partition into p groups, we then have a *double discriminant analysis* whose linear discriminant functions coincide with the factors of the correspondence analysis of the (p, q) contingency table cross-tabulating the two partitions.

The two submatrices now have analogous roles. We call them $\mathbf{Z}_1$ and $\mathbf{Z}_2$ and the whole matrix is called $\mathbf{Z}$, and no longer $\mathbf{R}$. Therefore we have

$$\mathbf{Z} = [\mathbf{Z}_1|\mathbf{Z}_2] \tag{90}$$

where the size of $\mathbf{Z}$ is $(n, p + q)$, the size of $\mathbf{Z}_1$ is (n, p), and the size of $\mathbf{Z}_2$ is (n, q).

Note that the linear subspaces $\mathscr{V}_{Z_1}$ and $\mathscr{V}_{Z_2}$ of $\mathbb{R}^n$ have in common the straight line generated by vector $\mathbf{e}_n$, all of whose components equal 1.

Let us define $\hat{\mathbf{Z}}_1$ and $\hat{\mathbf{Z}}_2$ by

$$\hat{\mathbf{Z}}_1 = \left(\mathbf{I} - \frac{1}{n}\mathbf{e}_n\mathbf{e}_n'\right)\mathbf{Z}_1 \tag{91}$$

$$\hat{\mathbf{Z}}_2 = \left(\mathbf{I} - \frac{1}{n}\mathbf{e}_n\mathbf{e}_n'\right)\mathbf{Z}_2 \tag{92}$$

$\hat{\mathbf{Z}}_1$ and $\hat{\mathbf{Z}}_2$ are the centered submatrices corresponding to $\mathbf{Z}_1$ and $\mathbf{Z}_2$.

It is equivalent to find two vectors $\hat{\mathbf{Z}}_1\mathbf{a}$ and $\hat{\mathbf{Z}}_2\mathbf{b}$ whose correlation is maximum, or two vectors $\mathbf{Z}_1\mathbf{a}$ and $\mathbf{Z}_2\mathbf{b}$ subject to the conditions

$$\mathbf{e}_n'\mathbf{Z}_1\mathbf{a} = \mathbf{e}_n'\mathbf{Z}_2\mathbf{b} = 0 \tag{93}$$

and having a maximum correlation ($\mathscr{V}_{Z_1}$, for example, is a direct sum of $\mathscr{V}_{\hat{Z}_1}$ and $\mathbf{e}_n$).

It is simple to determine that $\mathbf{Z}_1'\mathbf{Z}_1$ and $\mathbf{Z}_2'\mathbf{Z}_2$ are two diagonal (p, p) and (q, q) matrices whose diagonal elements are the group sizes, and that $\mathbf{Z}_1'\mathbf{Z}_2$ is the contingency table that cross-tabulates the two partitions.

Let us denote, using notation similar to that of Chapter II,

$$\mathbf{F} = \frac{1}{n}\mathbf{Z}_1'\mathbf{Z}_2 \qquad \mathbf{D}_p = \frac{1}{n}\mathbf{Z}_1'\mathbf{Z}_1 \qquad \mathbf{D}_q = \frac{1}{n}\mathbf{Z}_2'\mathbf{Z}_2 \tag{94}$$

Let us write the fundamental equation of discriminant analysis (70):

$$\mathbf{Z}_1'\mathbf{Z}_2\left(\mathbf{Z}_2'\mathbf{Z}_2\right)^{-1}\mathbf{Z}_2'\mathbf{Z}_1\mathbf{a} = \lambda\mathbf{Z}_1'\mathbf{Z}_1\mathbf{a} \tag{95}$$

Or, again,

$$\mathbf{D}_p^{-1}\mathbf{F}\mathbf{D}_q^{-1}\mathbf{F}'\mathbf{a} = \lambda\mathbf{a} \tag{96}$$

This equation shows that $\mathbf{a}$ is a factor of the correspondence analysis of matrix $\mathbf{F}$ of size (p, q), relative to the eigenvalue λ. The equation

$$\mathbf{e}_n'\mathbf{Z}_1\mathbf{a} = 0 \tag{97}$$

leads to

$$\sum_{j=1}^{p} f_{\cdot j}a_j = 0 \tag{98}$$

and signifies that the factors are centered.

Moreover, equations (15) and (16) of Section III.1.2 are written here as

$$\mathbf{F}\mathbf{b} = \beta\mathbf{D}_p\mathbf{a}$$

$$\mathbf{F}'\mathbf{a} = \beta\mathbf{D}_q\mathbf{b} \tag{99}$$

The reader will recognize the "transition equations" of Section II.2.3 with $\beta = \sqrt{\lambda}$.

This allows us to establish directly the fact that the eigenvalues extracted in correspondence analysis are *less than or equal to* 1. Furthermore, they can be interpreted in terms of the *discriminating power* of the factors with respect to partitions under consideration.

CHAPTER IV

Multiple Correspondence Analysis

Two-way correspondence analysis, which we described in Chapter II, can be readily generalized to encompass more than two sets of categories. The method of *multiple* correspondence analysis, which we present here, is a simple extension of two-way analysis. It is characterized by straightforward computations, interesting properties, and simple rules for interpreting the resulting maps.

The principles for this method, which were outlined in 1972 (Benzecri, 1972; Lebart and Tabard, 1973), actually stem from the work of the statistician C. Burt (1950). Other types of extensions have been proposed by Benzecri (1964), Escofier-Cordier (1965), and, more recently, by Masson (1974) (whose work is based on work done by Horst, 1961; Carroll 1968; and Kettenring 1971). See also Hayashi (1950), McKeon (1966), and Gifi (1981). Here we base our presentation on the one published by Lebart and Tabard in 1973.

IV.1. DEFINITIONS AND NOTATION

Survey data generally includes a number of responses to questions in completely disjunctive form: this means that the various response categories are mutually exclusive, and that only one category is chosen.

The k response categories to a given question allow us to partition the sample into k groups (at the most).

Example 1. The question, "How old are you?" may have eight response categories:

1. Less than 25 years.	5. From 40 to 44 years.
2. From 25 to 29 years.	6. From 45 to 49 years.
3. From 30 to 34 years.	7. 50 years or more.
4. From 35 to 39 years.	8. Don't know/no answer.

Example 2. If the question is "Do you own a dishwasher?" the answers are:

1. Yes.   2. No.

These two questions in complete disjunctive form lead to two ways of classifying all the individuals in the sample. The analysis of the correspondence matrix that cross-tabulates the two classifications may be generalized to the case of Q classifications where Q is an integer greater than 2.

IV.1.1. Notation

The number of questions is called Q. A single question q consists of a set p_q of response categories. When the two are not likely to be confused, both the set and the number of its elements are designated by the same letter.

The total number of response categories, p, contained in the questionnaire is

$$p = \sum_{q=1}^{Q} p_q \qquad (1)$$

The number of individuals who responded to questionnaire Q is called n.

H is the set whose elements consist of all the series of Q categories, each of which is taken from a different question. The elements of H therefore comprise the sum-total of possible responses by the subjects.

Each element of H corresponds to one cell of the multiway contingency table cross-tabulating the Q questions. We must note, however, that this hypertable is in general almost empty. If 1000 individuals are given 12 questions, each of which has 10 categories, $n = 1000$, whereas the number of elements of H is 10^{12}. Thus, at the most, only one cell out of one billion will be nonempty.

We denote by $\mathbf{Z}$ the matrix with n rows and p columns describing the response of the n individuals with binary coding.

Matrix $\mathbf{Z}$ is the juxtaposition of Q submatrices (see Figure 1):

$$\mathbf{Z} = \left[\mathbf{Z}_1, \mathbf{Z}_2, \ldots, \mathbf{Z}_q, \ldots, \mathbf{Z}_Q \right] \tag{2}$$

Submatrix $\mathbf{Z}_q$ (with n rows and p_q columns) is such that its ith row contains $p_q - 1$ times the value zero, and once the value 1, in the column corresponding to the category of question q chosen by subject i. In other words, matrix $\mathbf{Z}_q$ describes the partition of the n individuals that is created by the responses to question q. This type of matrix was previously discussed in Section III.2.3 with regard to discriminant analysis and in Section III.2.5.

Finally, we define a matrix $\mathbf{R}$ with n rows and Q columns. $\mathbf{R}$ is the *condensed coding matrix* of $\mathbf{Z}$. Cell (i, q) contains the number r_{iq} of the category of question q chosen by subject i. It is obvious that $r_{iq} \leq p_q$ (see Figure 1).

The computational programs use matrix $\mathbf{R}$ as input data, thus reducing considerably the volume of calculations (cf. Chapter VI).

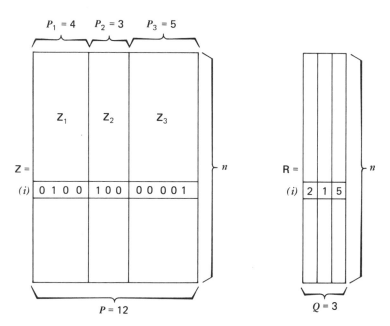

Figure 1.   Example of homologous matrices $\mathbf{Z}$ and $\mathbf{R}$.

IV.1.2. Burt's Table Associated with Z

Let us consider matrix Z such that $Z = [Z_1, Z_2, \ldots, Z_Q]$. The square matrix

$$B = Z'Z \tag{3}$$

is called Burt's *contingency table* associated with Z, the matrix of responses.

Matrix B is made up of Q^2 blocks.

The qth square matrix $Z_q'Z_q$ is a (p_q, p_q) *diagonal* matrix, since two categories of one question cannot be chosen simultaneously.

The block $(Z_q'Z_{q'})$ whose index is (q, q') is the *contingency table* that cross-tabulates the responses to the two questions q and q'.

The diagonal matrix D is a (p, p) matrix that has the same diagonal elements as B; these diagonal elements are the frequencies for each of the categories.

We may also consider matrix D as consisting of Q^2 blocks. Only the Q diagonal blocks are nonzero matrices; the qth diagonal block such that $D_q = Z_q'Z_q$ is the diagonal matrix whose diagonal terms are the frequencies corresponding to the various categories of question q.

IV.2. TWO QUESTIONS (TWO-WAY CORRESPONDENCE)

The response matrix Z is written $Z = [Z_1|Z_2]$.

It is then equivalent, from the point of view of *describing* the relationships among categories, to performing any of the following:

1. A correspondence analysis of the (n, p) matrix Z.
2. A correspondence analysis of the (p, p) matrix B.
3. A correspondence analysis of the (p_1, p_2) matrix $Z_1'Z_2$.
4. A canonical analysis of the two column blocks Z_1 and Z_2 (which is in this case a double discriminant analysis).

Note that the equivalence between analyses 3 and 4 was established previously in Section III.2.5.

IV.2.1. First Equivalence (Between 1 and 2)

Let us show that analyses 1 and 2 provide the same factors of norm 1. The αth factor Φ_α extracted from analysis 1 is such that

$$\frac{1}{Q}D^{-1}Z'Z\Phi_\alpha = \mu_\alpha\Phi_\alpha \tag{4}$$

since, using the notation we adopted for correspondence analysis, the matrix on the left-hand side of this equation would be written

$$\mathbf{D}_p^{-1}\mathbf{F'D}_n^{-1}\mathbf{F} \tag{5}$$

and here

$$\mathbf{F} = \frac{1}{nQ}\mathbf{Z}$$

$$\mathbf{D}_p = \frac{1}{nQ}\mathbf{D} \tag{7}$$

$$\mathbf{D}_n = \frac{1}{n}\mathbf{I}_n \tag{8}$$

(where $\mathbf{I}_n$ is the (n, n) identity matrix). We recognize equation (4) in the formula

$$\mathbf{D}_p^{-1}\mathbf{F'D}_n^{-1}\mathbf{F\Phi}_\alpha = \mu_\alpha\mathbf{\Phi}_\alpha \tag{9}$$

Let us consider matrix $\mathbf{B}$ such that $\mathbf{B} = \mathbf{Z'Z}$. $\mathbf{B}$ is *symmetric*. Its row and column marginals are the diagonal elements of the matrix $Q\mathbf{D}$.
In analyzing $\mathbf{B}$, we have a new matrix $\mathbf{F}$:

$$\mathbf{F} = \frac{1}{nQ^2}\mathbf{B} \tag{10}$$

And the corresponding new matrices $\mathbf{D}_n$ and $\mathbf{D}_p$ are equal:

$$\mathbf{D}_p = \mathbf{D}_n = \frac{1}{nQ}\mathbf{D} \tag{11}$$

Therefore, the matrix that is to be diagonalized in this case is written

$$\frac{1}{Q^2}\mathbf{D}^{-1}\mathbf{B'D}^{-1}\mathbf{B} \tag{12}$$

When we premultiply the two members of equation (4) by $(1/Q)\mathbf{D}^{-1}\mathbf{B}$, we immediately obtain

$$\frac{1}{Q^2}\mathbf{D}^{-1}\mathbf{BD}^{-1}\mathbf{B\Phi}_\alpha = \mu_\alpha^2\mathbf{\Phi}_\alpha \tag{13}$$

The factors are therefore identical for both analyses.

IV.2.2. Second Equivalence (Between 1 and 3)

Let us now show that, for each pair of factors $(\varphi_\alpha, \psi_\alpha)$ relative to the same eigenvalue λ_α extracted from the analysis of the contingency table $Z_1' Z_2$, there is a corresponding factor Φ_α from analyzing Z (or B) such that:

$$\Phi_\alpha = \begin{bmatrix} \varphi_\alpha \\ \psi_\alpha \end{bmatrix} \qquad (14)$$

Recall that we have the notation $D_1 = Z_1' Z_1$ and $D_2 = Z_2' Z_2$ and that

$$D = \begin{bmatrix} D_1 & 0 \\ 0 & D_2 \end{bmatrix} \qquad (15)$$

The diagonal elements of D_1 and D_2 are the row and column marginals of the matrix $Z_1' Z_2$.

Analysis of this table leads us to the *double transition* equations

$$\varphi_\alpha = \frac{1}{\sqrt{\lambda_\alpha}} D_1^{-1} Z_1' Z_2 \psi_\alpha \qquad (16)$$

$$\psi_\alpha = \frac{1}{\sqrt{\lambda_\alpha}} D_2^{-1} Z_2' Z_1 \varphi_\alpha \qquad (17)$$

These equations can be written as a system of equations:

$$D_1^{-1}(D_1 \varphi_\alpha + Z_1' Z_2 \psi_\alpha) = \left(1 + \sqrt{\lambda_\alpha}\right) \varphi_\alpha \qquad (18)$$

$$D_2^{-1}(D_2 \psi_\alpha + Z_2' Z_1 \varphi_\alpha) = \left(1 + \sqrt{\lambda_\alpha}\right) \psi_\alpha \qquad (19)$$

$$\begin{bmatrix} D_1 & 0 \\ 0 & D_2 \end{bmatrix}^{-1} \begin{bmatrix} D_1 & Z_1' Z_2 \\ Z_2' Z_1 & D_2 \end{bmatrix} \begin{bmatrix} \varphi_\alpha \\ \psi_\alpha \end{bmatrix} = \left(1 + \sqrt{\lambda_\alpha}\right) \begin{bmatrix} \varphi_\alpha \\ \psi_\alpha \end{bmatrix} \qquad (20)$$

This equation is written in a more condensed form after multiplying its two members by $\frac{1}{2}$ (that is, $1/Q$):

$$\frac{1}{Q} D^{-1} Z' Z \Phi_\alpha = \left(\frac{1 + \sqrt{\lambda_\alpha}}{2}\right) \Phi_\alpha \qquad (21)$$

The reader will recognize equation (4) with

$$\mu_\alpha = \frac{1 + \sqrt{\lambda_\alpha}}{2} \qquad (22)$$

If λ_α is the αth largest eigenvalue extracted from the analysis of $Z_1'Z_2$, then equation (22) gives the αth largest eigenvalue of the analysis of Z.

If, for example, $p_1 \leq p_2$, the analysis of Z leads to the following results:

1. p_1 factors of the type $\begin{bmatrix} \varphi_\alpha \\ \psi_\alpha \end{bmatrix}$, corresponding to the eigenvalue $(1 + \sqrt{\lambda_\alpha})/2$.

2. p_1 factors of the type $\begin{bmatrix} \varphi_\alpha \\ -\psi_\alpha \end{bmatrix}$, corresponding to the eigenvalue $(1 - \sqrt{\lambda_\alpha})/2$.

3. $p_2 - p_1$ factors of the type $\begin{bmatrix} 0 \\ \xi_\alpha \end{bmatrix}$ corresponding to the eigenvalue $\frac{1}{2}$ (the ξ_α axes complete the basis of ψ_α in $\mathbb{R}^n$).

The results of the three equivalent analyses are shown in Table 1.

Note 1.   In the analysis of the disjunctive matrix Z, the points that represent the various response categories of the two questions are elements of the same set, which is the set of Z's columns.

On the other hand, in analyzing the contingency table $Z_1'Z_2$, they are split up into row-points and column-points. The fact that the maps obtained in the space of the first factors are identical (although they are dilated, since the eigenvalues are not the same) shows that it is valid in some sense to *simultaneously* represent the row-points and the column-points in correspondence analysis.

In Section II.2.5, we have already demonstrated the optimal nature of this representation.

Note 2.   These three analyses, based on the same raw data, give similar results, but with different eigenvalues, and therefore different proportions of explained variances. The relationships among the explained variances (see

Table 1.

Analyzed Table	Dimension	Factor	Norm of Factor	Eigenvalue
$Z_1'Z_2$ contingency table	(p_1, p_2)	φ in $\mathbb{R}^{P_1}$, ψ in $\mathbb{R}^{P_2}$	$\varphi'D_1\varphi = n$, $\psi'D_2\psi = n$	λ
$Z = [Z_1|Z_2]$ disjunctive table	(n, p) with $p = p_1 + p_2$	$\Phi = \begin{bmatrix} \varphi \\ \psi \end{bmatrix}$	$\Phi'D\Phi = nQ$	$\mu = \dfrac{1 + \sqrt{\lambda}}{2}$
$B = Z'Z$ Burt's table	(p, p)	Φ	$\Phi'D\Phi = nQ$	μ^2

Table 1) demonstrate that they are always much higher when a contingency table $Z_1'Z_2$ is analyzed than when a matrix Z is analyzed.

Generally, the analysis of disjunctive coding matrices always results in small explained variances, giving a far too conservative view of the proportion of extracted information (see also Chapter VII).

Thus the sum of the nontrivial eigenvalues extracted from the analysis of Z is

$$\frac{p_1 + p_2}{2} - 1 \tag{23}$$

Since the eigenvalues are less than or equal to 1, no one factor can have a percentage of explained variance greater than

$$\frac{2 \times 100}{p_1 + p_2 - 2} \tag{24}$$

Let us take, for example, the $(26, 17)$ contingency table described in Section II.4. The percentages of explained variance obtained for the two first axes are 34.1% and 19.6%, whereas the analysis of the $(3267, 53)$ binary table Z, which leads to the same eigenvectors, produces percentages of 3.4% and 3.1%. In this latter analysis, the trace has the value $\frac{53}{2} - 1 = 25.5$, which provides an upper bound of 3.9% for the largest eigenvalue.

IV.3. GENERALIZATION TO MORE THAN TWO QUESTIONS

Generalization to the problem of more than two questions requires preliminary reformulation of the two-way problem.

The matrix $Z = [Z_1, Z_2, \ldots, Z_q, \ldots, Z_Q]$ has p columns, to which there correspond p points of $\mathbb{R}^n$. Let us consider the space $\mathbb{R}^n$. Each submatrix Z_q generates a linear subspace $\mathscr{V}_q$ with p_q dimensions.

All these linear subspaces have in common at least the first bisector (the vector all of whose components are equal to one). The rank of matrix Z is therefore at the most equal to

$$p - (Q - 1) \tag{25}$$

Let φ_q be the vector whose p_q components are the coordinates of a point m_q of $\mathscr{V}_q$ in the basis defined by the columns of Z_q.

The coordinates of m_q in $\mathbb{R}_n$ are the components of $m_q = Z_q\varphi_q$.

The square of the distance of this point m_q to the origin, or its Euclidean distance, is

$$\varphi_q' Z_q' Z_q \varphi_q = \varphi_q' D_q \varphi_q \tag{26}$$

The *correspondence analysis* of the contingency table that cross-tabulates two questions, q and q', is reduced to studying the relative positions of the subspaces $\mathcal{V}_q$ and $\mathcal{V}_{q'}$. As we have seen, this leads to a *canonical analysis* of the matrix $[\mathbf{Z}_q | \mathbf{Z}_{q'}]$.

The double transition equations, (16) and (17), are written (we have omitted the index α in order to simplify the notation)

$$\varphi_q = \frac{1}{\sqrt{\lambda}} \mathbf{D}_q^{-1} \mathbf{Z}_q' \mathbf{Z}_{q'} \varphi_{q'} \tag{27}$$

$$\varphi_{q'} = \frac{1}{\sqrt{\lambda}} \mathbf{D}_{q'}^{-1} \mathbf{Z}_{q'}' \mathbf{Z}_q \varphi_q \tag{28}$$

From these equations we can deduce the following:

$$\mathbf{Z}_q \varphi_q = \frac{1}{\sqrt{\lambda}} \mathbf{Z}_q \mathbf{D}_q^{-1} \mathbf{Z}_q' \mathbf{Z}_{q'} \varphi_{q'} \tag{29}$$

$$\mathbf{Z}_{q'} \varphi_{q'} = \frac{1}{\sqrt{\lambda}} \mathbf{Z}_{q'} \mathbf{D}_{q'}^{-1} \mathbf{Z}_{q'}' \mathbf{Z}_q \varphi_q \tag{30}$$

namely,

$$\mathbf{m}_q = \frac{1}{\sqrt{\lambda}} \mathbf{P}_q \mathbf{m}_{q'} \tag{31}$$

$$\mathbf{m}_{q'} = \frac{1}{\sqrt{\lambda}} \mathbf{P}_{q'} \mathbf{m}_q \tag{32}$$

where

$$\mathbf{P}_q = \mathbf{Z}_q (\mathbf{Z}_q' \mathbf{Z}_q)^{-1} \mathbf{Z}_q' \tag{33}$$

and

$$\mathbf{P}_{q'} = \mathbf{Z}_{q'} (\mathbf{Z}_{q'}' \mathbf{Z}_{q'})^{-1} \mathbf{Z}_{q'}' \tag{34}$$

The matrices $\mathbf{P}_q$ and $\mathbf{P}_{q'}$ represent the projection operators on $\mathcal{V}_q$ and $\mathcal{V}_{q'}$ (the projections are illustrated on Figure 2).

Equations (31) and (32) state that the orthogonal projection of $\mathbf{m}_q$ on $\mathcal{V}_{q'}$ is collinear to $\mathbf{m}_{q'}$ (and likewise for $\mathbf{m}_{q'}$ on $\mathcal{V}_q$).

When canonical analysis is presented as a problem of finding the smallest angles between two mappings, it does not lend itself to generalization to the problem of more than two questions.

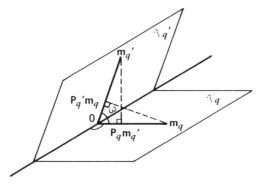

Figure 2

However, the canonical analysis of $[\mathbf{Z}_q | \mathbf{Z}_{q'}]$ can be formulated in the following way.

Find two points $\mathbf{m}_q$ and $\mathbf{m}_{q'}$ such that the mean of the sum of their squared distances to the origin is constant:

$$\boldsymbol{\varphi}_q' \mathbf{D}_q \boldsymbol{\varphi}_q + \boldsymbol{\varphi}_{q'}' \mathbf{D}_{q'} \boldsymbol{\varphi}_{q'} = 2n \qquad (35)$$

and such that the distance of the point $\mathbf{m} = \mathbf{m}_q + \mathbf{m}_{q'}$ to the origin is maximum.

The square of this distance is

$$\|\mathbf{m}\|^2 = \boldsymbol{\varphi}_q' \mathbf{D}_q \boldsymbol{\varphi}_q + \boldsymbol{\varphi}_{q'}' \mathbf{D}_{q'} \boldsymbol{\varphi}_{q'} + 2\boldsymbol{\varphi}_q' \mathbf{Z}_q' \mathbf{Z}_{q'} \boldsymbol{\varphi}_{q'} \qquad (36)$$

namely,

$$\|\mathbf{m}\|^2 = 2n\left(1 + \frac{1}{n}\boldsymbol{\varphi}_q' \mathbf{Z}_q' \mathbf{Z}_{q'} \boldsymbol{\varphi}_{q'}\right) \qquad (37)$$

Maximizing $\|\mathbf{m}\|^2$ with the constraint (35) or with the two constraints

$$\boldsymbol{\varphi}_q' \mathbf{D}_q \boldsymbol{\varphi}_q = \boldsymbol{\varphi}_{q'}' \mathbf{D}_{q'} \boldsymbol{\varphi}_{q'} = n \qquad (38)$$

leads to the same result. As we saw in Section III.1.2, the two Lagrange multipliers relative to the last two constraints are equal.

With the single constraint, the problem is readily generalized to more than two questions.

Let us designate by $\boldsymbol{\varphi}_1, \boldsymbol{\varphi}_2, \ldots, \boldsymbol{\varphi}_Q$, respectively, the vectors of the components of Q points, $\mathbf{m}_1, \mathbf{m}_2, \ldots, \mathbf{m}_q$ in the bases $\mathbf{Z}_1, \mathbf{Z}_2, \ldots, \mathbf{Z}_q$; and let

$\mathbf{m} = \mathbf{m}_1 + \mathbf{m}_2 + \cdots + \mathbf{m}_Q$. The quantity to be maximized is

$$\|\mathbf{m}\|^2 = \sum \left\{ \varphi_q' \mathbf{Z}_q' \mathbf{Z}_{q'} \varphi_{q'} | q \in Q; q' \in Q \right\} \tag{39}$$

with the constraint

$$\sum \left\{ \varphi_q' \mathbf{D}_q \varphi_q | q \in Q \right\} = Qn \tag{40}$$

If $\boldsymbol{\Phi}$ is the vector with p components defined by

$$\boldsymbol{\Phi}' = \left\{ \varphi_1', \varphi_2', \ldots, \varphi_Q' \right\} \tag{41}$$

the problem becomes maximizing:

$$\boldsymbol{\Phi}'\mathbf{B}\boldsymbol{\Phi} \tag{42}$$

with the constraint

$$\boldsymbol{\Phi}'\mathbf{D}\boldsymbol{\Phi} = Qn \tag{43}$$

The required $\boldsymbol{\Phi}$ factors are therefore the eigenvectors of

$$\mathbf{D}^{-1}\mathbf{B} \tag{44}$$

relative to the largest eigenvalues. They are proportional to those extracted from the correspondence analysis of matrix $\mathbf{Z}$ and coincide, as we have seen, with the factors extracted from analyzing matrix $\mathbf{B}$, itself considered a data table.

Thus a correspondence analysis of the disjunctive matrix $\mathbf{Z}$, using a classical correspondence analysis program, can provide us with the expected results. However, this is only possible with small matrices, since the volume of computations rapidly increases with the size of the matrix. For example, with 30 questions, each of which has 10 response categories, the matrix that has to be diagonalized is $(300, 300)$. Fortunately, the particular structure of the matrix allows us to use computational procedures that eliminate certain technical obstacles, and reduces considerably the computational burden (cf. Chapter V).

IV.4. PROPERTIES OF MULTIPLE ANALYSES

Recall that the $\boldsymbol{\Phi}$ factors extracted from the analysis of matrix $\mathbf{Z}$ are such that

$$\frac{1}{Q}\mathbf{D}^{-1}\mathbf{B}\boldsymbol{\Phi} = \mu\boldsymbol{\Phi} \tag{45}$$

By rearranging the terms of this equation to show the components φ_q of $\boldsymbol{\Phi}$ relative to question q, as well as the blocks of matrices $\mathbf{D}$ and $\mathbf{B}$, we obtain

$$\frac{1}{Q} \sum \{\mathbf{D}_{q'}^{-1} \mathbf{Z}_{q'}' \mathbf{Z}_q \varphi_q | q \in Q\} = \mu \varphi_{q'} \qquad (46)$$

IV.4.1. φ_q Components Are Centered

The Q subsets of points corresponding to the p_q categories of a question q have the same center of gravity, which is therefore also the center of gravity of the entire set of points. J_q denotes the subset of the p values of the index j corresponding to the question q (J_q has p_q elements).

The coordinates of the subsets of points relative to question q are the columns of

$$\mathbf{Z}_q \mathbf{D}_q^{-1} \qquad (47)$$

and the diagonal elements of $(1/n)\mathbf{D}_q$ are the relative *masses* of the p_q points of subset q.

The ith coordinate of the center of gravity G_q is

$$g_{qi} = \sum_{j \in J_q} \frac{d_{jj}}{n} \frac{z_{ij}}{d_{jj}} = \frac{1}{n} \qquad \left(\text{since } \sum_{j \in J_q} z_{ij} = 1\right) \qquad (48)$$

Thus g_{qi} is independent of q: $g_{qi} = g_i$

The φ_q components corresponding to nontrivial factors are thus *centered* since these factors correspond to an analysis of the set of points after translation of the origin in G.

IV.4.2. Proportion of Variance Due to One Question and to One Category

Equation (46) demonstrated that the *total variance*, the sum of the nontrivial eigenvalues, is equal to

$$\frac{p}{Q} - 1$$

In particular, this trace is equal to 1 when all the questions have two response categories (when $p = 2Q$).

The square of the distance from the center of gravity of a category-point j is written, in $\mathbb{R}^n$,

$$d^2(G, j) = n \sum_{i=1}^{n} \left(\frac{z_{ij}}{d_{jj}} - \frac{1}{n}\right)^2 \qquad (49)$$

namely, taking into account the equation

$$\sum_{i=1}^{n} z_{ij} = d_{jj} \tag{50}$$

$$d^2(G, j) = n\left(\frac{1}{d_{jj}} - \frac{1}{n}\right) \tag{51}$$

The contribution to the total variance of category j is then

$$c(j) = \frac{d_{jj}}{nQ} d^2(G, j) = \frac{1}{Q}\left(1 - \frac{d_{jj}}{n}\right) \tag{52}$$

The contribution of question q to the total variance is

$$C(q) = \sum \{c(j)|j \in J_q\} = \frac{1}{Q}(p_q - 1) \tag{53}$$

Thus the *proportion of variance due to one question* is an increasing function of the number of response categories. The minimum proportion, $1/Q$, is obtained when a question has only two categories. We can check that

$$\sum_{q=1}^{Q} C(q) = \frac{p}{Q} - 1 \qquad \text{(total variance)} \tag{54}$$

The *proportion of variance due to one response category* increases as the number of responses in that category decreases. The maximum of $1/Q$ would be reached when a category has a zero response. Thus it is important to avoid categories with low response rates.

IV.4.3. Dimensionality of the Configuration of the p Categories in $\mathbb{R}^n$

The coordinates of the categories in $\mathbb{R}^n$ are the columns of $\mathbf{ZD}^{-1}$. They generate a subspace whose dimension is the rank of $\mathbf{ZD}^{-1}$, and therefore the rank of $\mathbf{Z} = [\mathbf{Z}_1, \mathbf{Z}_2, \ldots, \mathbf{Z}_Q]$.

All the $\mathscr{V}_q$ subspaces generated by the columns of the $\mathbf{Z}_q$ matrices have in common the first bisector Δ; therefore the maximum rank of $\mathbf{Z}$ is

$$p_1 + (p_2 - 1) + \cdots + (p_Q - 1) = p - Q + 1 \tag{55}$$

Therefore the maximum rank of $\mathbf{D}^{-1}\mathbf{Z}'\mathbf{Z}$, the matrix to be diagonalized, is $p - Q + 1$. But in the analysis with respect to the origin O, the first bisector Δ is the eigenvector corresponding to the eigenvalue 1 (the points are contained in the subspace which is $\mathbf{D}^{-1}$-orthogonal to Δ); in the analysis with respect to the center of gravity G, $(p - Q)$ nonzero eigenvalues are found.

Thus by choosing a basis *in the subspace of the set of points*, the problem can be reduced to finding eigenvalues and eigenvectors for a $(p - Q)$ matrix.

IV.4.4. Best Simultaneous Representation

The presentation of correspondence analysis that was outlined in Section II.2.5 may be given a specific formulation here thanks to the specific coding of matrix $\mathbf{Z}$.

We are seeking the abscissas of the n individuals and the p categories on the same axis such that:

1. The abscissa of an individual i is the arithmetic mean of his or her responses (after a dilation, which is kept to a minimum).
2. The abscissa of a category j is the arithmetic mean of the abscissas of the individuals who chose it (with the same dilation).

Of course, we obtain, as we did previously, the equations of *double transition* associated with the analysis of $\mathbf{Z}$:

$$\hat{\mathbf{\Psi}} = \frac{1}{\sqrt{\mu}} \frac{1}{Q} \mathbf{Z} \hat{\mathbf{\Phi}} \tag{56}$$

$$\hat{\mathbf{\Phi}} = \frac{1}{\sqrt{\mu}} \mathbf{D}^{-1} \mathbf{Z}' \hat{\mathbf{\Psi}} \tag{57}$$

where $\hat{\mathbf{\Psi}}_i$ is the abscissa of individual i, and $\hat{\mathbf{\Phi}}_j$ that of category j.

Introducing the other axes of the analysis on the basis of this particularly simple presentation is straightforward.

IV.4.5. Confidence Interval of a Supplementary Category

The abscissa $\hat{\mathbf{\Phi}}_{\alpha j}$ of category j on an axis α (whether the category is active or supplementary) is the product by the coefficient $1/\sqrt{\mu_\alpha}$ of the *arithmetic mean* of the $\hat{\mathbf{\Psi}}_{\alpha i}$ abscissas of the individuals who chose this response category. Recall this notation:

$$\hat{\mathbf{\Psi}}_{\alpha i} = \mathbf{\Psi}_{\alpha i} \sqrt{\mu_\alpha} \qquad \text{and} \qquad \hat{\mathbf{\Phi}}_{\alpha j} = \mathbf{\Phi}_{\alpha j} \sqrt{\mu_\alpha} \tag{58}$$

This suggests the following hypothesis test.

Suppose that a supplementary category j contains n_j individuals. The null hypothesis is that the n_j individuals are chosen at random (without replacement) among the n individuals in the analysis.

Under these circumstances the abscissa $\hat{\Phi}_{\alpha j}$ of a supplementary category concerning n_j individuals is a random variable, the product of $1/\sqrt{\mu_\alpha}$ and of the mean of n_j components drawn at random from the finite set of the n values of $\hat{\Psi}_{\alpha i}$. Therefore we have

$$E(\hat{\Phi}_{\alpha j}) = E(\hat{\Psi}_{\alpha i}) = 0 \tag{59}$$

and

$$\operatorname{var}(\hat{\Phi}_{\alpha j}) = \frac{n - n_j}{n - 1} \cdot \frac{\operatorname{var}(\hat{\Psi}_{\alpha i})}{n_j \mu_\alpha} = \frac{n - n_j}{n - 1} \cdot \frac{1}{n_j} \tag{60}$$

If the projections of the points on axis α are distributed according to the *normal law*, then we can say that approximately 95% of the projections of the supplementary categories must lie between

$$-2\sqrt{\frac{n - n_j}{(n - 1)n_j}} \quad \text{and} \quad +2\sqrt{\frac{n - n_j}{(n - 1)n_j}}$$

Even if these projections are not normally distributed, this interval offers a convenient frame of reference, thanks to the central limit theorem. In particular, the quantities

$$\hat{\Phi}_{\alpha j}\sqrt{n_j \frac{n - 1}{n - n_j}} \tag{61}$$

have the same standard deviation. They allow us to compare respective significances of several supplementary categories. These quantities are named "test-values."

IV.5. TWO SPECIAL CASES

When all the questions have two response categories, the multiple analysis is reduced to a principal components analysis of the questions characterized by only one of their categories. In another case, the question set can be partitioned into two groups within which the questions are independent. Then the multiple analysis is reduced to the analysis of a submatrix of Burt's table.

IV.5.1. All the Questions Have Two Categories (Cazes, 1976; Nakhle, 1976)

In this case, although the reduction discussed in Section IV.4.3 may be applied with a substantial saving of computation time, we can instead obtain the matrix to be diagonalized directly. This matrix is the variables' correlation matrix, where the variables are represented by only one of their categories ($p - Q = p/2$).

Let us develop equation (45) from above, where **D** is the diagonal matrix that has the same diagonal elements as **B**.

$$\frac{1}{Q} \sum_{j \in J} \frac{b_{ij}}{b_{ii}} \Phi_j = \mu \Phi_i \tag{62}$$

The category set J is partitioned into two subsets, J_1 and J_2, containing, respectively, the first and second categories of each of the Q questions:

$$J = J^1 \cup J^2 \tag{63}$$

For every q:

$$J_q = \left\{ J_q^1, J_q^2 \right\} \text{ with } J_q^1 \in J^1 \text{ and } J_q^2 \in J^2. \tag{64}$$

Note the following equations, for every $q \in Q$:

$$b_{ij_q^1} + b_{ij_q^2} = b_{ii}, \qquad i \in J \tag{65}$$

$$b_{j_q^1 j_q^2} + b_{j_q^2 j_q^2} = n, \quad b_{j_q^1 j_q^1} \Phi_{j_q^1} = -b_{j_q^2 j_q^2} \Phi_{j_q^2} \tag{66}$$

Therefore we can limit the summation of equation (62) to the set J^1 alone, and henceforth note its elements as j, without index:

$$\frac{1}{Qb_{ii}} \sum \left\{ b_{ij} \Phi_j - \frac{(b_{ii} - b_{ij}) b_{jj} \Phi_j}{(n - b_{jj})} \middle| j \in J^1 \right\} = \mu \Phi_i \tag{67}$$

which can be written

$$\sum \left\{ \frac{n b_{ij} - b_{ii} b_{jj}}{Q(n - b_{jj}) b_{ii}} \Phi_j \middle| j \in J^1 \right\} = \mu \Phi_i \tag{68}$$

Let us calculate the empirical second order centered moments of the Q variables characterized by their first categories:

$$\text{cov}(i, j) = \frac{1}{n}\left(b_{ij} - \frac{b_{ii}b_{jj}}{n}\right) \tag{69}$$

$$\text{var}(j) = \frac{1}{n}\left(b_{jj} - \frac{b_{jj}^2}{n}\right) \tag{70}$$

The general term of the correlation matrix of the Q variables is written

$$C_{ij} = \frac{nb_{ij} - b_{ii}b_{jj}}{\left[(n - b_{jj})b_{jj}(n - b_{ii})b_{ii}\right]^{1/2}} \tag{71}$$

Clearly, if (Φ, μ) is the solution of equation (68), then (Φ^*, μ^*) is the solution of

$$\sum\left\{C_{ij}\Phi_j^* | j \in J^1\right\} = \mu^*\Phi_j^* \tag{72}$$

with

$$\Phi_j = \Phi_j^* \frac{(n - b_{jj})^{1/2}}{b_{jj}^{1/2}} \tag{73}$$

and

$$\mu^* = \mu Q \tag{74}$$

IV.5.2. The Case Where Multiple Analysis Is Reduced to a Two-Way Analysis

The case of a two-way correspondence is particularly interesting from a computational standpoint. The analysis of a (p, p) Burt's table is equivalent to the correspondence analysis of a contingency table cross-tabulating the categories of the two questions, which leads us to diagonalizing a matrix whose dimension is determined by the smaller of the two numbers p_1 and p_2.

We note the following property, which is useful for applications. If the question set Q is partitioned into two subsets, Q_1 and Q_2, within which the questions are independent, then the analysis of the Q questions reduces itself to a two-way correspondence analysis.

We state here that two questions q and q' are independent if the joint frequency is the product of the corresponding margins

$$\mathbf{Z}'_q \mathbf{Z}_{q'} = \frac{1}{n} \mathbf{d}_q \mathbf{d}'_{q'} \tag{75}$$

where components of the vectors $\mathbf{d}_q$ and $\mathbf{d}_{q'}$ are, respectively, the diagonal elements of $\mathbf{Z}'_q \mathbf{Z}_q$ and $\mathbf{Z}'_{q'} \mathbf{Z}_{q'}$ (that is, the diagonal elements of $\mathbf{D}_q$ and $\mathbf{D}_{q'}$ by definition of these matrices).

Let us rewrite equation (45) by partitioning $\mathbf{\Phi}$ into two blocks $\mathbf{\Phi}_{Q_1}$ and $\mathbf{\Phi}_{Q_2}$; let us section matrices $\mathbf{B}$ and $\mathbf{D}$ also into four blocks, so as to bring out the dichotomy of $Q = Q_1 + Q_2$:

$$\mathbf{B} = \begin{bmatrix} \mathbf{B}_{11} & \mathbf{B}_{12} \\ \mathbf{B}_{21} & \mathbf{B}_{22} \end{bmatrix} \qquad \mathbf{D} = \begin{bmatrix} \mathbf{D}_1 & 0 \\ 0 & \mathbf{D}_2 \end{bmatrix} \tag{76}$$

We obtain these two equations:

$$\frac{1}{Q}\left(\mathbf{D}_1^{-1}\mathbf{B}_{11}\mathbf{\Phi}_{Q_1} + \mathbf{D}_1^{-1}\mathbf{B}_{12}\mathbf{\Phi}_{Q_2} \right) = \mu \mathbf{\Phi}_{Q_1} \tag{77}$$

$$\frac{1}{Q}\left(\mathbf{D}_2^{-1}\mathbf{B}_{21}\mathbf{\Phi}_{Q_1} + \mathbf{D}_2^{-1}\mathbf{B}_{22}\mathbf{\Phi}_{Q_2} \right) = \mu \mathbf{\Phi}_{Q_2} \tag{78}$$

Note that the Q_1 (respectively Q_2) diagonal blocks of $\mathbf{D}_1^{-1}\mathbf{B}_{11}$ (respectively $\mathbf{D}_2^{-1}\mathbf{B}_{22}$) are unitary matrices.

$$q \in Q_k; \quad q' \in Q_k; \quad q = q' \Rightarrow \mathbf{D}_q^{-1}\mathbf{Z}'_q\mathbf{Z}_q = \mathbf{I}_q \quad (k \in \{1, 2\}) \tag{79}$$

On the other hand, we have, for $k \in \{1, 2\}$,

$$q \in Q_k; \quad q' \in Q_k; \quad q \ne q' \Rightarrow \mathbf{D}_q^{-1}\mathbf{Z}'_q\mathbf{Z}_{q'} = \frac{1}{n}\mathbf{D}_q^{-1}\mathbf{d}_q\mathbf{d}'_{q'} \tag{80}$$

If we designate by $\mathbf{e}_q$ a vector whose q components are 1,

$$\mathbf{D}_q^{-1}\mathbf{Z}'_q\mathbf{Z}_{q'} = \frac{1}{n}\mathbf{e}_q\mathbf{d}'_{q'} \tag{81}$$

Finally, the equations $\mathbf{d}'_{q'}\mathbf{\Phi}_{q'} = 0$ imply that, for $k \in \{1, 2\}$,

$$\mathbf{D}_k^{-1}\mathbf{B}_{kk}\mathbf{\Phi}_{Q_k} = \mathbf{\Phi}_{Q_k} \tag{82}$$

The above system ((77) and (78)) is then written

$$\mathbf{D}_1^{-1}\mathbf{B}_{12}\Phi_{Q_2} = (\mu Q - 1)\Phi_{Q_1} \qquad (83)$$

$$\mathbf{D}_2^{-1}\mathbf{B}_{21}\Phi_{Q_1} = (\mu Q - 1)\Phi_{Q_2} \qquad (84)$$

from which, by substitution,

$$\mathbf{D}_2^{-1}\mathbf{B}_{21}\mathbf{D}_1^{-1}\mathbf{B}_{12}\Phi_{Q_2} = (\mu Q - 1)^2\Phi_{Q_2} \qquad (85)$$

Thus Φ_{Q_2} is obtained by diagonalizing a (Q_2, Q_2) matrix. It is then easy to deduce Φ_{Q_1}.

Note that $\mathbf{B}_{12}$ is obtained by *juxtaposition of the contingency tables* that cross-tabulate the category set of the questions of the first group with those relative to the second group. (For further work on the juxtaposition of several contingency tables, the reader should consult Leclerc, 1975, and Cazes, 1976).

IV.6.   EXAMPLE: SURVEY PROCESSING

The first part of this section briefly recalls the steps involved in processing survey data, and demonstrates the shortcomings in blindly producing masses of tables.

The second part deals with a practical example, showing how multiple correspondence analysis is a tool for simultaneously analyzing a large number of related contingency tables and finding out which are relevant and which are redundant, as well as uncovering significant interactions.

IV.6.1.   Operational Survey Procedures

Variables that describe a statistical unit in a socioeconomic survey can be classified roughly into two main groups:

(a)   Demographic variables: These "basic" variables may be used to design the survey and are sometimes already known before the interview: for example, number of children, selected aspects of geographical location, and number of rooms in the home. They are essentially demographic or economic, or provide a general description of the social position of the family.

(b)   Variables related to the content of the survey, which may involve one or more main content areas. They can be divided into three subgroups that differ in level and quality of measurement:

1.   Variables that are derived from answers to factual questions (possession of water heater) and that must be distinguished from the first group

because they do not concern, for example, either a social group or other standard demographic category.

2. Variables that are also factual, describing the behavior of the respondent or of the other members of the household, but that result from ambiguous, inaccurate, or difficult-to-code answers (to such questions as "Do you watch T.V.?").

3. Attitudinal variables that play an important role in the understanding and prediction of socioeconomic phenomena. These variables, however, yield only tenuous and unreliable information, especially if they are studied separately.

Processing the survey usually requires using the demographic variables to design the contingency tables, in order to understand and explain the "behavior" of the variables resulting from the content of the survey.

Frequency distributions (e.g., average consumption of a given item by an occupational group) as well as cross-tabulations (repeating the one-way tabulation for each type of town, which is the same as crossing the "occupational group" and "type of town" variables) are assembled. The demographic categories, in a certain way, play the role of predictor variables and provide suggestions for explaining the phenomena. If the survey is not the first of its kind, the previous experience of statisticians in planning the tabulation procedures may be used to cover some of the problems the survey is supposed to answer. If the survey concerns a previously unexplored field, the cross-tabulations will probably be partly redundant and yet will prove insufficient. The number of tabulations that are assumed necessary a priori may be considerable. Furthermore, the sequential generation of cross-tabulations does not take into account the relations between the elements of the tabulation design. Thus if one studies the time spent on leisure activities by occupational group, then by level of education, and finally by level of income, the interrelations between the three criteria have not been taken into account. In order to do as complete an analysis as possible, it is therefore necessary to use the demographic variables comprehensively, taking into account their interrelations, in order to avoid wasting time when analyzing the results. Multiple correspondence analysis can be useful in this situation.

IV.6.2. An Example of "Predictive Map"

We use, for this example, a socioeconomic sample survey relating to the "Living conditions and aspirations of the French" (see Lebart and Houzel van Effentere, 1980). The subsample of 1000 used here is representative of the population over 18 years.

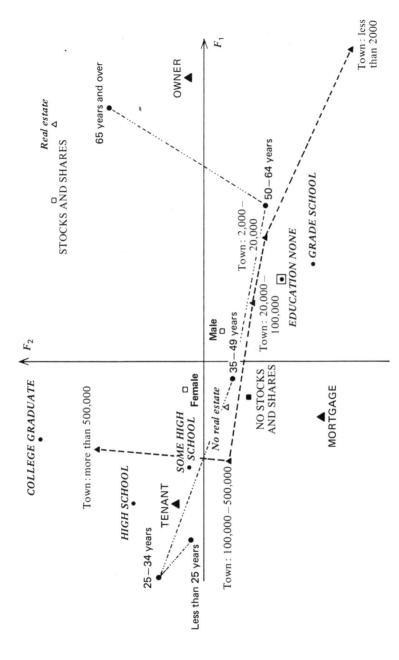

Figure 3

101

The first step in the analysis consists of constructing a predictive map, derived from multiple correspondence analysis of a set of demographic variables (objective characteristics).

Figure 3 summarizes the first phase of the analysis, the design of a demographic chart; we limit ourselves here to the first two principal axes in order to demonstrate the part of the procedure that can easily be visualized. But the automatic part of the analysis, in particular the selection of the most important supplementary variables, can of course be performed in the space of the k first axes that are thought to be more relevant (see Section IV.6.2.e).

Figure 3 shows the relationships between these objective characteristics, as described by the first two axes of the analysis.

The set of active variables contains 7 questions, containing a total of 25 categories:

1. Sex.
2. Level of education (five categories).
3. Home Ownership (four categories).
4. Ownership of stocks and shares (two categories).
5. Ownership of real estate (two categories).
6. Age (five categories).
7. Size of town (five categories).

Computer printouts of the detailed dictionary, Burt's contingency table, and the histogram of eigenvalues are shown in Tables 2, 3, and 4.

(a) *Overall Interpretation of Figure 3.*   This figure is more an intermediate tool than a result in itself: the main step is the projection of supplementary elements (here: variables related to the content of the survey) onto this two-dimensional space.

However, we can comment on the observed patterns. Polygonal lines have been drawn to join the ordered items of the variables "age" and "size of town." These lines are unfolded approximately along the horizontal axis, which thus describes the relationship between the age of the population and the level of urbanization in the country under study.

Home ownership is related to both age and degree of urbanization. The ownership of stocks and shares, which involves only a very small proportion of the sample (12%), is related to age, but in an upward direction also related to the educational level. However, the highest levels of education are positioned on the upper left side of the map, that is to say a rather young and urban part of the sample.

(b) *Visualized Regression and Supplementary Variables.*   The 1000 respondents (the rows of matrix $\mathbf{Z}$) could also have appeared on Figure 3.

Table 2. Dictionary of variables and response categories

```
VARIABLES OF TYPE  1           7

CORRESPONDING CATEGORIES      25
```

```
Q1- SEX OF THE RESPONDENT
   /  SEXM=MALE             /  SEXF=FEMALE          /

Q2- WHAT IS THE HIGHEST LEVEL OF EDUCATION ATTAINED?
   /  EDU0=NONE             /  EDU1=GRADE SCHOOL    /  EDU2=SOME HIGH SCHOOL    /  EDU3=HIGH SCHOOL GRAD.  /
   /  EDU4=SOME COLLEGE     /

Q10- DO YOU RENT OR OWN YOUR LODGING ?
   /  LOD1=MORTGAGE         /  LOD2=OWNER           /  LOD3=TENANT              /  LOD4=RENT FREE          /

Q19 - HAVE YOU GOT ANY STOCKS AND SHARES ?
   /  STO1=YES              /  STO2=NO              /

Q20 - DO YOU OWN REAL ESTATE ?
   /  HOU1=YES              /  HOU2=NO              /

Q28 - AGE OF THE RESPONDENT ?
   /  AGE1=UNDER 25 YEARS   /  AGE2=25 - 34 YEARS   /  AGE3=35 - 49 YEARS       /  AGE4=50 - 64 YEARS      /
   /  AGE5=65 YEARS AND OVER /

Q33 - SIZE OF TOWN (NUMBER OF INHABITANTS)
   /  SIZ1=LESS THAN 2000   /  SIZ2=2000 - 20000    /  SIZ3=50,000 - 100,000    /  SIZ4=MORE THAN 100000   /
   /  SIZ5=MORE THAN 500000 /
```

```
I  TOTAL NUMBER OF   ROWS                 1000 I
```

103

Table 3.   Burt's table

```
        SEXM  SEXF  EDU0  EDU1  EDU2  EDU3  EDU5  LOD1  LOD2  LOD3  LOD4  STO1  STO2  HOU1  HOU2  AGE1  AGE2  AGE3  AGE4  AGE5  SIZ1  SIZ2  SIZ3  SIZ4  SIZ5

SEXM I  469.
SEXF I    0.  531.

EDU0 I  102.   94.  196.
EDU1 I  164.  163.    0.  327.
EDU2 I   69.   91.    0.    0.  160.
EDU3 I   65.  102.    0.    0.    0.  167.
EDU5 I   69.   81.    0.    0.    0.    0.  150.

LOD1 I   62.   58.   17.   45.   16.   27.   15.  120.
LOD2 I  151.  142.   59.  116.   37.   39.   42.    0.  293.
LOD3 I  224.  302.  105.  151.   98.   87.   85.    0.    0.  526.
LOD4 I   32.   29.   15.   15.    9.   14.    8.    0.    0.    0.   61.

STO1 I   54.   67.   12.   26.   19.   27.   37.   11.   60.   45.    5.  121.
STO2 I  415.  464.  184.  301.  141.  140.  113.  109.  233.  481.   56.    0.  879.

HOU1 I   35.   47.   11.   21.   14.   17.   19.    7.   48.   23.    3.   39.   43.   82.
HOU2 I  434.  484.  185.  306.  146.  150.  131.  113.  245.  503.   58.   82.  836.    0.  918.

AGE1 I   18.   22.    9.   14.   16.   16.    1.    1.    9.   28.    2.    2.   38.    3.   37.   40.
AGE2 I  111.  136.   27.   47.   62.   62.   56.   23.   24.  180.   20.   18.  229.   11.  236.    0.  247.
AGE3 I  169.  187.   54.   60.   63.   63.   54.   68.   91.  182.   15.   35.  321.   27.  329.    0.    0.  356.
AGE4 I   84.  104.   47.   27.   16.   16.   20.   20.   80.   79.    9.   27.  161.   20.  168.    0.    0.    0.  188.
AGE5 I   87.   82.   59.   13.   13.   10.    6.    8.   89.   57.   15.   39.  130.   21.  148.    0.    0.    0.    0.  169.

SIZ1 I   42.   41.   19.   43.   12.    8.    1.    7.   59.   11.    6.    4.   79.    7.   76.    2.   13.   31.   25.   12.   83.
SIZ2 I   40.   47.   17.   34.   13.   15.    8.   20.   33.   29.    5.    9.   78.   12.   75.    4.   16.   34.   20.   13.    0.   87.
SIZ3 I   81.   94.   35.   67.   34.   21.   18.   27.   62.   80.    6.   22.  153.   12.  163.    8.   34.   71.   26.   36.    0.    0.  175.
SIZ4 I  161.  168.   70.  110.   52.   57.   40.   48.   72.  191.   18.   36.  293.   25.  304.   15.  102.  100.   60.   52.    0.    0.    0.  329.
SIZ5 I  145.  181.   55.   73.   49.   66.   83.   18.   67.  215.   26.   50.  276.   26.  300.   11.   82.  120.   57.   56.    0.    0.    0.    0.  326.

        SEXM  SEXF  EDU0  EDU1  EDU2  EDU3  EDU5  LOD1  LOD2  LOD3  LOD4  STO1  STO2  HOU1  HOU2  AGE1  AGE2  AGE3  AGE4  AGE5  SIZ1  SIZ2  SIZ3  SIZ4  SIZ5
```

Table 4. Eigenvalues

SUM OF THE EIGENVALUES   2.57141352

HISTOGRAM OF THE FIRST EIGENVALUES

	EIGENVALUE	PERCENTAGE	PERCENTAGE CUM.	
1	0.25001878	9.72	9.72	**
2	0.21836281	8.49	18.21	***
3	0.18465650	7.18	25.40	**
4	0.17460611	6.79	32.19	***
5	0.16103518	6.26	38.45	***
6	0.15802696	6.15	44.59	***
7	0.15417504	6.00	50.59	**
8	0.14570984	5.67	56.26	***
9	0.14088601	5.48	61.74	***
10	0.13440627	5.23	66.96	***
11	0.12898514	5.02	71.98	**
12	0.12528747	4.87	76.85	**
13	0.12234123	4.76	81.61	***************************************
14	0.11767256	4.58	86.18	**************************************
15	0.10597984	4.12	90.31	**********************************
16	0.09571891	3.72	94.03	******************************
17	0.08693265	3.38	97.41	**************************
18	0.06661423	2.59	100.00	*********************

Under this type of coding, the transition relations are interpreted simply: one respondent is characterized by seven answers; to obtain its position one simply locates the center of gravity of the corresponding seven point-answers on Figure 3, the coordinates being dilated by $1/\sqrt{\lambda_\alpha}$ relative to each axis.

In a similar fashion, projecting the answer to a supplementary question is equivalent to calculating the center of gravity of the respondents involved in each response, and expanding them as above.

This procedure is clearly related to multiple regression, being a descriptive variant of it. The responses to the demographic variables may indeed be considered as exogenous or independent variables creating a certain linear subspace. However, instead of projecting the supplementary answers (dependent or explained variables) directly onto this subspace, we begin by fitting it with a subspace of smaller dimensions (two dimensions on Figure 3). Also, instead of studying the regression coefficients (coordinates in the basis of the demographic variables), we simply observe the relative positions of the different responses.

Figure 3, therefore, constitutes a lattice on which the different "weave designs" are placed, according to the themes of the survey.

From a computational point of view, it is much less expensive to project the response categories of a supplementary question than to cross-tabulate the question with all the demographic variables. Moreover looking at the map is much easier and more meaningful than reading the corresponding tables.

Furthermore, it is not necessary to look at all of the supplementary responses, since they may be extremely numerous (several thousand in the survey we are examining). The answers that have the most significant positions (on the first k axes for instance) can be selected automatically. Supposing an answer concerns n individuals, the null hypothesis is: n points, in the space of the first k axes, are taken at random among the 1000 points referenced by their coordinates. It is easy under these conditions to construct significance levels for the distance of the additional question from the origin. One may then either keep the answers corresponding to a given significance level or classify them in increasing order of significance (see Section IV.4.5).

Of course, as this automatic filter is relatively inexpensive, the supplementary questions can be crossed two by two, in order to detect certain types of interactions.

(c) The Projection of Supplementary Variables. A selection of supplementary variables is plotted in Figure 4. This selection can be done by the computer program, according to a level of significance (test values), taking

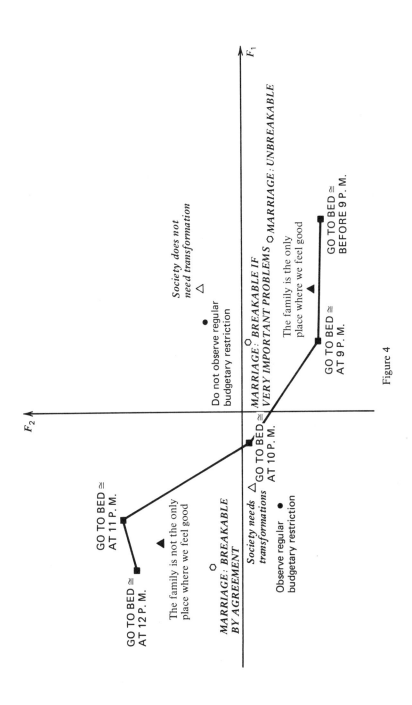

Figure 4

107

into account the frequency of a category and the distance of the corresponding point to the origin of the axes.

Figure 4 could have been drawn on a transparency, in order to merge it easily with Figure 3.

However, we must remember that the variables of Figure 3 play a quite different role. In this example, the set of supplementary variables deals with opinions, attitudes, or items describing certain aspects of the behavior of the respondents.

Comparing Figures 3 and 4 will show, at a glance, the objective context into which the behavioral items are inserted.

In the right part of Figure 4 one finds rather conservative opinions about family, marriage, society, while the left part is characterized by "modernist" items, with a higher level of dissatisfaction.

The "hours at which people go to bed" has a monotonic relationship with the first axis. This indicator is linked simultaneously with age, degree of urbanization, and educational level. It could provide an indication of how opinions are shared.

As a matter of fact, in the original application, both sets of active and supplementary variables were much larger. The heuristic power of the method lies in the possibility of handling hundreds of supplementary variables (thousands of categories), and then automatically selecting the most relevant items, with respect to the chosen descriptive frame.

(d)   The Reciprocal Approach.   In the previous section we developed a framework based on objective variables, and then projected attitudinal variables into this framework. This could have been done in reverse. It is advisable, in fact, to create a framework based on attitudinal variables, and then project the objective (demographic) variables into this context. Any adequate analysis of a data set requires that both be done.

Each type of analysis develops a specific point of view.

(e)   Exploring More than Two Dimensions: The Utility of Clustering Techniques.   If we analyze the procedure that we have sketched briefly above, we discover that we are comparing different areas of the two-dimensional space. We seldom treat the space as continuous (i.e., look for underlying dimensions of the space), except when we interpret trajectories of ordered items. Part of this process can be done automatically, by clustering the set of individuals (not represented on Figures 3 and 4, but underlying the whole operation, since variable points are dilated centroids of individuals), and describing the clusters in terms of categories of the other variables, demographic or attitudinal. This point of view is explored in Chapter V.

CHAPTER V

Automatic Classification— Clustering Techniques Used with Principal Axes Methods

Principal components analysis, correspondence analysis, and multiple correspondence analysis rarely provide an exhaustive analysis of a set of data. It is sometimes useful to perform a preliminary clustering of the observations because it will help reduce the complexity of the analysis, when we come to use one of the above techniques. Additionally, it is of value to use classification analysis to summarize the configuration of points obtained from a principal axis analysis. In other words, a further reduction in the dimensionality of the data is valuable and leads to results that are easier to analyze.

The automatic classification techniques presented in this chapter do not make any claim to optimality. Nevertheless, they give relatively fast, economical, and easily interpretable results. They comprise a two-step algorithm, using both partitioning and hierarchical clustering.

V.1. INTRODUCTION—CLASSIFICATION AND PRINCIPAL AXES METHODS

Automatic classification techniques are used to group objects or individuals described by a number of variables or characteristics. This form of data analysis is extremely popular at the moment, and publications in this field are numerous.

Although the field is too new to be able to give a complete summary of the methods that have been developed, much of the work has been covered, for example, in the following references: Friedman and Rubin (1967), Cormack (1971), Anderberg (1973), Benzecri (1973), Sneath and Sokal (1973), Hartigan (1975), and Gordon (1981).

The reader should look at the above literature in order to obtain a fundamental grasp of the concept of classification. Here, we limit ourselves to considerations that are likely to interest users of classification techniques.

First of all, we attempt to answer two questions that should be asked about any multivariate descriptive statistical analysis (MDSA) technique.

QUESTION 1.   Under what circumstances are these methods to be used?

QUESTION 2.   What types of results do they produce?

These methods are used in the same situation as descriptive principal axes analysis: the user is faced with a rectangular matrix of numbers. This matrix can be a contingency table (cross-tabulating two partitions of a population), a binary matrix (with values of 0 or 1 according to whether an individual or object has a certain characteristic or attribute), or a matrix of numerical values (value of variable j for individual i at the intersection of row i and column j of the matrix).

In any case, this matrix must satisfy certain classical requirements of data analysis. It must be *homogeneous* in its *form* and in its *content*. Computing and comparing distances between rows (the rows generally represent individuals or objects) and between columns must make sense. The matrix must be so *large* that its structure is not obvious to the naked eye, or easily revealed through elementary statistical manipulations. Finally, it must be *amorphous* as much as possible: there must not exist an a priori structure within its rows and columns, such as functional dependencies (see also Chapter VII).

The use of automatic classification techniques implies some basic underlying concepts with respect to the purpose of the analysis. Either it is assumed that certain groups must exist among the observations, or, on the contrary, the analysis requires a grouping of the observations. In other words, a two-dimensional, *continuous* visualization of the statistical relationships is not enough. There is also an implicit or explicit interest in uncovering *groups* of individuals or of characteristics.

This brings us to the second question, which is dependent on the first one, and which concerns the nature of the expected results. They will consist of either *partitions* of the data set (i.e., the rows or columns of the matrix), or *hierarchies of partitions*, which we ultimately define more precisely. Sometimes they will comprise *trees* in the sense of graph theory, that is, *threes whose nodes are the objects to classify.* Finally, the results might be *overlapping clusters*, or *dense areas* where many individuals or characteristics remain unclassified.

A given set of results might be reached through different steps and might lead to different interpretations. For example, the problem may be to discover a partition that really exists, and that was hypothesized before carrying out the statistical analysis, or one that revealed itself after the analysis. Conversely, it may be useful to employ the partitions as tools or as surrogates in the computations that make it easier to explore the data. This last case is a generalization of the histograms of one-dimensional statistics: in order to make the analysis easier, the observations are grouped into homogeneous groups, even if these groups imply a somewhat arbitrary division of a continuous space.

In any case, using principal axes methods in conjunction with classification makes it possible to identify groups and to determine their relative positions. Often partitions or tree structures are used to amplify the results of preliminary principal axes analysis during the exploratory phases of data analysis.

There are several families of classification algorithms: agglomerative algorithms, in which the clusters are built by successive pairwise agglomerations of objects, and which provide a hierarchy of partitions of the objects; divisive algorithms, which proceed by successive dichotomizations of the entire set of objects, and which can also provide a hierarchy of partitions; finally, algorithms leading directly to partitions, such as the methods of clustering around moving centers, or other minimum variance algorithms.

The method presented here uses, in its first phase, a partitioning technique that is designed for large data matrices (containing several hundred to several thousand observations) (see Section V.2).

The groups obtained from this phase are then clustered through an agglomerative algorithm (Section V.3).

This method combines the advantages of both approaches, namely:

1. The ability to treat very large matrices.
2. A detailed description of the main clusters.
3. The ability to make a critical choice with respect to the number of clusters.

V.2.   CLUSTERING AROUND MOVING CENTERS

Although it is based on a relatively *slight theoretical basis*, and its efficacy is largely attested to by empirical results, the method of classification around moving centers is probably the partitioning method that is best adapted to large data sets. It belongs to the class of methods known as *k-mean* algorithms. It is used as an adjunct to other methods (such as clustering prior to principal axes analysis), or directly as a descriptive tool.

This method is particularly well adapted to large numerical data sets, because the data is read directly: the data matrix is saved on auxiliary memory (such as a magnetic tape or disk) and is read several times in sequential fashion, without requiring large amounts of computer memory. Reading directly also takes advantages of the particular manner in which the data is coded, leading to appreciable economies in computation time in the case of binary coding.

V.2.1. Theoretical Basis of the Algorithm

The algorithm underlying the programs presented here can be attributed mainly to Forgy (1965), although numerous developments (sometimes earlier, Thorndike, 1953, most often later, Ball and Hall, 1967; MacQueen, 1967; Diday, 1971; Diday et al., 1980) arose independently.

Let us suppose we wish to partition a set I of n individuals characterized by p variables. Let us also suppose that the space $\mathbb{R}^p$ containing the n individual-points has an appropriate *Euclidean distance* denoted by d. The maximum number of groups desired is k_c.

STEP 0. k_c provisional group centers are determined. (In our programs, this is done by a pseudorandom selection, without replacement, of k_c individuals in the population to be clustered, as proposed by MacQueen.) The k_c centers,

$$C_1^0, C_2^0, \ldots, C_k^0, \ldots, C_{k_c}^0$$

immediately create a partition P^0 of I into k_c clusters,

$$I_1^0, I_2^0, \ldots, I_k^0, \ldots, I_{k_c}^0$$

The assignment rule is: an individual i belongs to I_k^0 if point i is nearer to C_k^0 than to all other centers. (The clusters are delimited in space by convex polyhedral divisions formed by the median planes of the segments joining all the center pairs.) See Figure 1.

STEP 1. k_c new cluster centers are determined:

$$C_1^1, C_2^1, \ldots, C_{k_c}^1$$

by using the centers of gravity of the clusters

$$I_1^0, I_2^0, \ldots, I_{k_c}^0$$

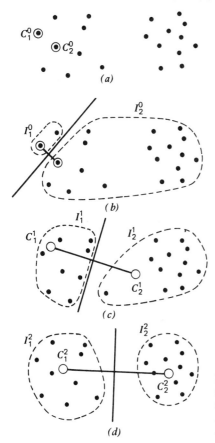

Figure 1. (a) Random drawing of centers C_1^0 and C_2^0. (b) Constitution of groups I_1^0 and I_2^0. (c) New centers C_1^1 and C_2^1; new groups I_1^1 and I_2^1. (d) New centers C_1^2 and C_2^2; new groups I_1^2 and I_2^2.

These new centers create a new partition P^1 constructed according of the same rule as for P^0. The partition P^1 is formed with the clusters denoted

$$I_1^1, I_2^1, \ldots, I_{k_c}^1$$

STEP m. k_c new cluster centers are determined:

$$C_1^m, C_2^m, \ldots, C_{k_c}^m$$

by taking the centers of gravity of the clusters

$$I_1^{m-1}, I_2^{m-1}, \ldots, I_{k_c}^{m-1}$$

These new centers create a new partition P^m of the set I, formed by the clusters

$$I_1^m, I_2^m, \ldots, I_{k_c}^m$$

The algorithm stops either when two succeeding iterations lead to the same partition, or when a conveniently chosen criterion (for example, the measure of the within-groups variance) stops decreasing significantly, or when a previously established maximum number of iterations is reached.

In every case, the resulting partition depends upon the initial choice of the centers at step 0.

V.2.2. Elementary Rationale for the Algorithm

We show that the within-group variance can only become smaller (or remain constant) between step m and step $m + 1$. *Assignment rules* (which are programming conventions specific to each variant or each specification of the algorithm) make it possible to force this decrease to be a strict one, and therefore to conclude that the algorithm does converge since the initial set I is finite (obviously it is not the convergence itself, but the *rate* of convergence that justifies using this method in practice).

Let us suppose that the n individuals of the set I to be clustered have relative weights p_i such that

$$\sum_{i=1}^{n} p_i = 1 \tag{1}$$

and let

$$d^2(i, C_k^m) \tag{2}$$

be the square of the distance between individual i and the center of group k at step m.

We are interested in the *criterion* quantity

$$v(m) = \sum_{k=1}^{k_c} \left\{ \sum_{i \in I_k^m} p_i d^2(i, C_k^m) \right\} \tag{3}$$

Recall that at step m, group I_k^m is formed from the individuals who are closer to C_k^m than to all the other centers (these centers are the centers of gravity of the groups I_k^{m-1} of the preceding step).

The within-group variance at step m is therefore the quantity

$$V(m) = \sum_{k=1}^{k_c} \left\{ \sum_{i \in I_k^m} p_i d^2(i, C_k^{m+1}) \right\} \tag{4}$$

where C_k^{m+1} is the center of gravity of group I_k^m. At step $m + 1$, the criterion quantity is written

$$v(m + 1) = \sum_{k=1}^{k_c} \left\{ \sum_{i \in I_k^{m+1}} p_i d^2\left(i, C_k^{m+1}\right) \right\} \tag{5}$$

We show that

$$v(m) \geq V(m) \geq v(m + 1) \tag{6}$$

which establishes the simultaneous decrease of the criterion and of the within-group variance.

We note

$$p(k) = \sum_{i \in I_k^m} p_i \tag{7}$$

First of all, let us note that

$$v(m) = V(m) + \sum_{k=1}^{k_c} p(k) d^2\left(C_k^{m+1}, C_k^m\right) \tag{8}$$

according to Huyghens' theorem, which establishes the first part of the inequality.

The second part follows from the fact that between the parentheses that appear in the definitions of $V(m)$ and $v(m + 1)$, only the assignments of the points to the centers change. Since I_k^{m+1} is the set of points that are closer to C_k^{m+1} than to all the other centers, the distances can only have decreased (or stayed the same) in the course of this reassignment.

V.2.3. Related Techniques

There exist many algorithms that have general principles similar to the algorithm of clustering around moving centers, but that differ in several respects. For more detailed and precise information about the techniques of clustering around moving centers, the reader is advised to consult, in addition to the publications already mentioned, the work of Benzecri, 1973 (Vol 1: Taxonomy, Chapter B-9), and Anderberg, 1973 (Chapter VII), as well as Diday et al., 1980.

(a) Thus in the technique of "dynamic clusters," the clusters are not characterized by a center of gravity, but by a certain number of individuals to be clustered, called "standards," who constitute a "nucleus" that has, for some applications, a better descriptive power than do point-centers. Assignment to clusters is done with the help of a "function of aggregation—separation." This formulation has led to several generalizations of the method (Diday, 1972).

(b) The so-called k-means method introduced by MacQueen (1967) starts with a pseudorandom selection of point-centers. However the rule for calculating the new centers is not the same. The position of the centers is modified before all of the individuals have been reassigned: each reassignment of individuals leads to a modification of the position of the corresponding center. This procedure therefore gives a high quality partition in a single iteration. But it depends on the order of the individuals on the data file, which was not the case for the technique discussed above.

(c) Other methods differ in the initial choice of the centers (equidistant individuals for Thorndike, 1953), in the introduction of *thresholds*, or in the introduction of *protection* whose purpose is to be able to modify the number of groups. Thus the technique proposed under the name of ISODATA by Ball and Hall (1965) uses several parameters to guide the building of the partition.

V.2.4. Stable Clusters

Algorithms for clustering around moving centers converge toward *local optima*. The problem of finding an *optimal* partition into k_c clusters (using within-group variance as a criterion, which must then be minimized over the set of possible partitions into k_c clusters) has not yielded, to date, a satisfactory algorithm. The partitions obtained generally depend on the initial centers chosen.

The procedure for finding *stable clusters*, proposed by Diday (1972), is at least a partial remedy for this situation. Its main advantage is that it elaborates the results obtained in the rigid framework of a single partition by highlighting high-density regions of the cluster of individual-points.

The technique consists in performing several partitions starting with several different sets of centers, and keeping as *stable clusters* the sets of individuals that are always assigned to the same cluster in each of the partitions. See Figure 2.

Let us designate by $P_1, P_2, \ldots, P_s$ the s partitions into k_c clusters (where each cluster may require a different number of iterations).

The cluster that is indexed by $(k_1, k_2, \ldots, k_s)$ contains the individuals that belonged first to cluster k_1 of P_1, then to cluster k_2 of $P_2, \ldots$, and finally to cluster k_s of P_s. This product partition thus contains $(k_c)^s$ clusters.

The nonempty classes of this partition constitute the stable clusters.

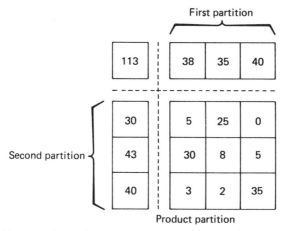

Figure 2. Stable groups in product partition (e.g., among the 38 individuals of group 1 of partition 1, 30 are found in group 2 of partition 2).

In practice the number of stable clusters with substantial bases is much smaller than $(k_c)^s$. For instance, it is not uncommon, after 4 partitions into 5 clusters have been performed on 1000 individuals, to have only 20 stable clusters with bases greater than 10 (even though the product partition contains $5^4 = 625$ clusters).

V.3. HIERARCHICAL CLASSIFICATION

V.3.1. Introduction

(a) Distances Between Elements and Between Groups. The general principles common to various hierarchical classification techniques are very simple.

1. At the outset we suppose that some measure of distance can be defined between pairs of objects that are to be classified. Sometimes this distance is simply a measure of dissimilarity. In this case, the triangular inequality, $d(x, y) \leq d(x, z) + d(x, z)$, is not required. The distances are not necessarily calculated at the outset: in this situation it must be possible to compute or recompute them from the coordinates of the object-points, and these coordinates must be easily accessible.

2. Then we suppose that there exist *rules for calculating the distances between disjoint clusters* of objects. This between cluster distance can generally be calculated directly from the distances of the various elements involved in the clustering.

For example, if x, y, and z are three objects, and if x and y are clustered in h, we can define the distance from this cluster to z as the smallest distance from the various elements of h to z:

$$d(h, z) = \min\{d(x, z), d(y, z)\} \tag{9}$$

This distance is called the *single linkage* (Sneath, 1957; Johnson, 1967), and it is discussed later in greater detail.

Another rule that is simple and frequently used is that of the *group average*: for two objects x and y clustered in h:

$$d(h, z) = \frac{d(x, z) + d(y, z)}{2} \tag{10}$$

Generally, if x and y designate disjoint subsets of the set of objects, containing, respectively, n_x and n_y elements, h is a subset containing $(n_x + n_y)$ elements, and we define

$$d(h, z) = \frac{n_x d(x, z) + n_y d(y, z)}{n_x + n_y} \tag{11}$$

(b)   Classification Algorithm.   The general concepts we describe here are difficult to trace historically. They are based on common sense rather than on formalized theory. The most systematic and earliest explanation is perhaps that of Lance and Williams (1967).

The basic algorithm for ascending hierarchical classification is as follows. We call *points* either the objects to be classified or the clusters of objects generated by the algorithm.

1.   At step 1 there are n points to classify (which are the n objects).
2.   We find the two points that are closest to one another and aggregate them into a new point.
3.   We calculate the distances between the new point and the remaining points. We return to step 1 with only $n - 1$ points to classify.
4.   We again find the two closest points and aggregate them; we calculate the new distances, and repeat the process until there is only one point remaining.

(c)   Terminology.   We make a few remarks to introduce the terminology associated with classification.

1.   In the case of single linkage, the algorithm uses distances in terms of the *inequalities* between them. The same tree could be obtained (with a

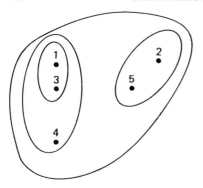

Figure 3

different scale) by simply using an ordering of pairs of objects. In this case the tree is drawn with equidistant levels.

2. The family of clusters built by such an ascending algorithm forms a so-called *hierarchy*. This family has the property of containing the entire set as well as every one of the objects taken separately. The other parts of this family are then either disjoint, or included in one another. Every time a new cluster is formed out of disjoint points, the new cluster becomes itself a new point, and therefore necessarily included in an ulterior cluster (Figure 3).

The hierarchy is an *indexed hierarchy* if a numerical value $v(h) \geq 0$ is associated with every part h of the hierarchy, such that the value is compatible with the inclusion relationship in the following way: if $h \subset h'$, then $v(h) \leq v(h')$. The hierarchy of Figure 4 is naturally indexed by the

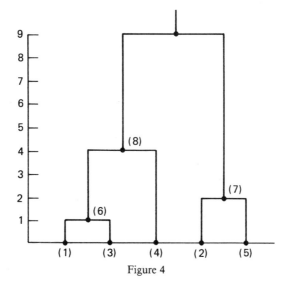

Figure 4

values of the distances corresponding to each aggregation step (these distances are plotted in the ordinate).

3. By "cutting" the tree of Figure 4 with a horizontal line, we obtain a *partition* in which the number of classes increases as the cut-point approaches the initial points.

A hierarchy allows us to obtain a set of *n* nested partitions containing 1 to *n* groups.

V.3.2. Single Linkage Classification and Minimum Spanning Tree

(a) Definition of an Ultrametric Distance. The method of classification presented above is very simple to perform, and has interesting properties that we formulate and discuss. We show that the concept of a hierarchy is closely linked to a class of distances between objects called *ultrametric distances.* For the hierarchy produced by the single linkage algorithm we show that the corresponding ultrametric distance is in a sense the closest to the original distance. This is called the *maximum lower ultrametric* distance, or the *subdominant ultrametric* distance. Then we show that applying this method is practically equivalent to solving a classical problem in operations research: finding the *minimum spanning tree* on a graph.

Recall that a set E is provided with distance d, if d is a positive mapping that satisfies the following conditions:

1. $d(x, y) = 0$ if and only if $x = y$.
2. $d(x, y) = d(y, x)$ (symmetry).
3. $d(x, y) \le d(x, z) + d(y, z)$ (triangular inequality).

' This distance is called an ultrametric distance if it satisfies the following condition, which is stronger than the triangular inequality:

4. $d(x, y) \le \max\{d(x, z), d(y, z)\}$.

(b) Equivalence Between an Ultrametric Distance and an Indexed Hierarchy. It is equivalent to provide a finite set E with an ultrametric distance or to define an indexed hierarchy of parts of this set.

Let us show first that *any indexed hierarchy makes it possible to define a distance between elements having the required properties.* For the distance $d(x, y)$ we take the value of the index corresponding to the smallest part that contains both x and y. Let us show in a general fashion that we always have:

$$d(x, y) \le \max\{d(x, z), d(y, z)\} \tag{12}$$

Recall that two parts of hierarchy H are either disjoint or linked by an inclusion relationship. Let us call $h(x, z)$ the smallest part of H that contains x and z (whose index is therefore $d(x, z)$). Since $h(x, z)$ and $h(y, z)$ are not disjoint, we have, for example, $h(x, z) \subset h(y, z)$. And since x, y, and z are all contained in $h(y, z)$, we necessarily have $h(x, y) \subset h(y, z)$; consequently $d(x, y) \leq d(y, z)$, which establishes the inequality.

Conversely, to every ultrametric distance d there can correspond an indexed hierarchy of which d is the associated index. We just have to apply the single linkage algorithm to the corresponding distance matrix. It is then seen that recomputing the distances at each step is unnecessary: instead one of the two aggregated points needs only to be deleted.

For example, if x and y are aggregated in t, the distances to the new point t should be calculated (Figure 5). But we necessarily have $d(z, x) \geq d(x, y)$ and $d(z, y) \geq d(x, y)$; otherwise (z, x) or (z, y) would have been aggregated instead of (x, y). For an ultrametric distance, this implies that $d(z, x) = d(z, y)$, which is expressed by saying that for an ultrametric distance, all of the triangles are isosceles triangles, the base being the smallest side. Let us prove this last result. We have:

$$d(z, x) \leq \max\{d(z, y), d(x, y)\}$$

therefore

$$d(z, x) \leq d(z, y) \tag{13}$$

Similarly,

$$d(z, y) \leq \max\{d(z, x), d(x, y)\}$$

therefore

$$d(z, y) \leq d(z, x) \tag{14}$$

It follows that $d(z, y) = d(z, x)$.

The computation of the distances from z to t is unnecessary because the two distances under consideration are equal.

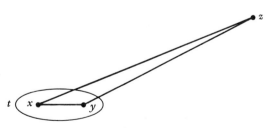

Figure 5

(c)   Subdominant Ultrametric Distance.   We have shifted from a metric distance to an ultrametric distance (that is, equivalently, to a hierarchy), by decreasing the values of certain distances. We can ask the following question: does there exist an ultrametric distance that is nearer (in a way undefined as yet) to the metric distance?

We can give the following partial answer. We say that metric distance d_1 is less than a metric distance d_2 if, for every x and y,

$$d_1(x, y) \le d_2(x, y) \tag{15}$$

(This definition makes it possible to assign a partial order relationship to the set of metric distances over a set E.)

The greatest ultrametric distance that is less than a metric distance d, in the preceding sense, is called the *maximum lower ultrametric distance* or *subdominant ultrametric distance*. This precisely is the one given by the single linkage algorithm. (For a complete proof, see Section V.7, the Appendix.)

(d)   Minimum Spanning Tree. Introduction and Definition.   The set of n points to be classified may be considered as a set of points in a space. This representation is classical if the objects are described by a series of p parameters: we have n points in the space $\mathbb{R}^p$. Then we can calculate a distance for every pair of points. More generally, if only the values of an index of dissimilarity are available (these do not necessarily have the properties of a distance), then we can represent the objects by points (of a plane for example), where every pair of objects is joined by a continuous line, to which is assigned the value of the index of dissimilarity.

Thus the set of objects and the values of the index are represented by a *complete valued graph*. But, if there are more than a few objects, this type of representation becomes messy. Then we attempt to extract a *partial graph* from this graph (with the same nodes and fewer edges); this partial graph is easier to represent, and permits us to summarize the values of the index.

Among partial graphs, those having a tree structure are particularly interesting, because they can be shown in two dimensions. (This type of tree is a tree in the sense of the theory of graphs; its nodes are the objects to classify. It is not to be confused with a tree made up of parts of a set, resulting from a classification technique.) A tree is *connected* (there is a path linking every pair of nodes), *without a cycle* (a cycle is a path leaving from and arriving at the same point without passing through the same edge twice). We can define in equivalent manner a tree with n nodes, either as a graph without a cycle having $(n - 1)$ edges or as a connected graph having $(n - 1)$ edges. The *length* of a tree is the sum of the "lengths" (values of the index) of its edges.

Among all the partial graphs that are trees, the *minimum spanning tree* has long held the attention of statisticians because of its excellent descriptive qualities, stemming from its relationship to hierarchical classification.

If, for example, we wish to rapidly detect, without a computer, the structural features of a correlation matrix of 30 variables, it is probably the easiest procedure to use.

First, we present the algorithms for finding a minimum spanning tree, then we show the equivalence with a single linkage classification. We suppose that all the edges of the graphs have different lengths (values of the index or of the distance), because under these circumstances the tree is unique and this simplifies the presentation of the algorithms.

Minimum Spanning Tree. Kruskal's Algorithm (1956).   The $n(n-1)/2$ edges are arranged in order of increasing values of the index. Starting with the first two edges, all of the edges that do not form a cycle with the edges already chosen are selected. The procedure is stopped when there are $n-1$ edges. In this way we are certain that we have obtained a tree (a graph without a cycle, having $n-1$ edges).

Minimum Spanning Tree. Prim's Algorithm (1957).   We begin with any node of the graph (see Figure 6). Step 1 consists of finding the nearest object v_1, that is, the shortest edge. Step k consists of joining to the existing series of edges V_{k-1} the shortest edge v_k that touches one of the vertices of V_{k-1} and that does not form a cycle with the edges of V_{k-1}. The tree obtained is of minimal length because V_k is at all times a minimal length tree over the k nodes.

Minimum Spanning Tree. Florek et al.'s Algorithm (1951).   At the first step, each node is joined to its nearest neighbor. This is equivalent to taking

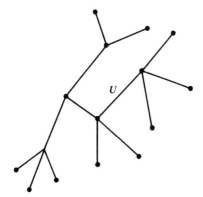

Figure 6.   Drawing of tree U.

the smallest distance in each row of the distance matrix. This rapid operation produces more than $n/2$ edges (or $(n - 1)/2$ if n is odd). It gives a *forest* F_1 (a family of trees, or simply, a graph without a cycle).

At step k, each tree of forest F_{k-1} (each connected component of the graph without a cycle) is joined to its nearest neighbor by taking as a distance between trees the shortest edge between any node of one and any node of the other. This process stops as soon as graph F_k is connected.

This algorithm is the fastest one to compute manually on rather large distance matrices. Generally there are only two or three steps.

Relationship Between Minimum Spanning Tree and Single Linkage (Gower and Ross, 1969).   Let V be a minimum spanning tree built from a distance matrix between n objects. Since V does not have a cycle, and is connected, there exists one and only one path joining two vertices x and y. Let us call $d_v(x, y)$ the length of the largest edge encountered on this path. We show that $d_v(x, y)$ is $d^*(x, y)$, the ultrametric distance of the smallest maximal jump between x and y (see Section V.7, the Appendix).

Let v be the largest edge between x and y. Suppressing v leads to dividing V into two separate connected components. If there exists a path from x to y (that does not necessarily go through the edges of V) whose largest edge is shorter than v, then there exists an edge u that is distinct from v, and shorter than v, joining the two connected components. Replacing v by u would give a tree of shorter length than that of V, which contradicts the definition of V. Thus $d_v(x, y)$, the length of v, is indeed the smallest maximal jump.

This argument provides a way of building the hierarchy associated with the single linkage starting with the minimum spanning tree V. This descending construction works in the following manner. We break the largest edge of V; we obtain thus the two groups that are farthest apart, and the

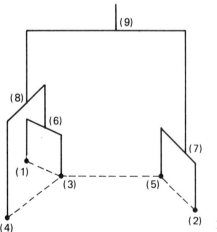

Figure 7

index corresponding to their fusion is the length of this edge. Then we break the edges in succession, in order of decreasing magnitude, descending along the hierarchy until the terminal elements, the objects themselves, are reached. The last broken edge corresponds to the two objects that were aggregated first in the ascending algorithm.

We can simultaneously represent the hierarchy and the minimum spanning tree in perspective as shown in Figure 7.

Some supplementary information is added to the classical dendrogram. Specifically, the relative positions of the points are better represented. For the practitioner of principal axes methods, it is often interesting to place the minimal length tree on the factorial planes, in order to remedy the deformations associated with projections.

V.3.3. Minimum Variance Algorithms and Related Techniques (see Ward, 1963)

(a) *Introduction and Notation.* The classification techniques discussed in the previous section have the advantage of being simple to compute, and they also have interesting mathematical properties.

For some applications the results can, however, be criticized. Specifically, single linkage has the defect of producing "chain effects." Thus for the points of Figure 8, groups (*a*) and (*b*) are never discernible in the hierarchical tree; additionally, the new nodes linking them are aggregated at the lowest level.

Other techniques for calculating distances may give more reliable results. We have mentioned the *group average distance*. The techniques of *minimum variance clustering* seek to optimize, at every step, the partition obtained by aggregating two elements, using criteria that are linked to variance. These techniques are particularly easy to compute when the clustering is performed after a *principal axes analysis*, and the objects to classify are located by their coordinates on the first axes of the analysis.

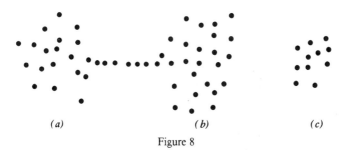

(*a*)   (*b*)   (*c*)

Figure 8

Here we consider the n objects to classify as points in a Euclidean space with p dimensions. Each point x_i (vector with p components) has a mass m_i. The entire mass of the set of points is noted $M = \sum_i m_i$. The square of the distance between the points x_i and x_j is noted:

$$\|x_i - x_j\|^2 = d^2(x_i, x_j) \tag{16}$$

The total variance of the set of points is the quantity

$$I = \sum_i m_i \|x_i - G\|^2 \tag{17}$$

where G is the center of gravity of the points

$$G = \frac{\sum_i m_i x_i}{M} \tag{18}$$

If there exists a partition of the set of objects into Q clusters, the qth cluster having a center of gravity G_q and a mass m_q, Huyghens' equation provides a decomposition of the quantity I into within cluster and between cluster variances according to the formula

$$I = \sum_q m_q \|G_q - G\|^2 + \sum_q \sum_{i \in q} m_i \|x_i - G_q\|^2 \tag{19}$$

(b) Reduction in Variance by Aggregation of Two Elements (Ward's Method).   Let x_i and x_j be two elements of mass m_i and m_j that are aggregated into one element x of mass $m = m_i + m_j$ with

$$x = \frac{m_i x_i + m_j x_j}{m_i + m_j} \tag{20}$$

We can decompose the variance I_{ij} of x_i and x_j with respect to G according to Huyghens' equation:

$$I_{ij} = m_i \|x_i - x\|^2 + m_j \|x_j - x\|^2 + m\|x - G\|^2 \tag{21}$$

Only the last term remains if x_i and x_j are replaced by their center of gravity x. The reduction in variance ΔI_{ij} is therefore

$$\Delta I_{ij} = m_i \|x_i - x\|^2 + m_j \|x_j - x\|^2 \tag{22}$$

By replacing $\mathbf{x}$ with its value as a function of $\mathbf{x}_i$ and $\mathbf{x}_j$, we obtain

$$\Delta I_{ij} = \frac{m_i m_j}{m_i + m_j}\|\mathbf{x}_i - \mathbf{x}_j\|^2 = \frac{m_i m_j}{m_i + m_j} d^2(\mathbf{x}_i, \mathbf{x}_j) \tag{23}$$

Thus the strategy of clustering based on the *minimum variance criterion* is as follows: instead of finding the two closest elements, we find the elements $\mathbf{x}_i$ and $\mathbf{x}_j$ corresponding to a *minimal* ΔI_{ij}—which is the same as considering the ΔI_{ij} as new dissimilarity indices. (By means of this transformation, points with small weights are more readily clustered.)

If we work with the coordinates of the points, we calculate the centers of gravity ($\mathbf{x}$ for $\mathbf{x}_i$ and $\mathbf{x}_j$). On the other hand, if we work with the distances, it is convenient to be able to calculate the new distances from the old (as was the case in the preceding techniques). The square of the distances between a point $\mathbf{z}$ and a cluster center $\mathbf{x}$ is written, as a function of the distances to $\mathbf{x}_i$ and $\mathbf{x}_j$,

$$d^2(\mathbf{x}, \mathbf{z}) = \frac{1}{m_i + m_j}\left\{ m_i d^2(\mathbf{x}_i, \mathbf{z}) + m_j d^2(\mathbf{x}_j, \mathbf{z}) - \frac{m_i m_j}{m_i + m_j} d^2(\mathbf{x}_i, \mathbf{x}_j)\right\}$$

$$\tag{24}$$

This formula is established by decomposing the variance of the doublet $(\mathbf{x}_i, \mathbf{x}_j)$ with respect to $\mathbf{z}$ into variance with respect to $\mathbf{x}$, and into variance of $\mathbf{x}$ with respect to $\mathbf{z}$ (Figure 9):

$$m_i\|\mathbf{x}_i - \mathbf{z}\|^2 + m_j\|\mathbf{x}_j - \mathbf{z}\|^2 = \underbrace{(m_i + m_j)\|\mathbf{x} - \mathbf{z}\|^2} + \underbrace{\frac{m_i m_j}{m_i + m_j}\|\mathbf{x}_i - \mathbf{x}_j\|^2}$$

$$\tag{25}$$

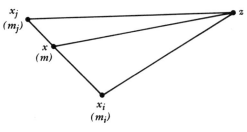

Figure 9

The expression of $d^2(\mathbf{x}, \mathbf{z})$ is immediately deduced from this. The process is repeated on the remaining elements and the new element is built by aggregation.

There are variants of this method (minimal variance loss) that use slightly different computational formulas. For example, we can find clusters having a minimal internal variance.

These techniques are discussed in Benzecri (1973).

V.4. RECIPROCAL NEIGHBORS—CHAIN SEARCH ALGORITHM

The main difficulty in building a hierarchical tree is the volume of computations. The basic algorithm (V.3.1), where the nodes are constructed one by one at each step of the algorithm, requires a number of calculations and of distance comparisons on the order of n^3 if there are n points to classify.

New algorithms greatly reduce the number of operations. The volume of calculations can decrease from n^3 to n^2, allowing the classification of several thousand points in a reasonable amount of time.

They utilize the concept of *reciprocal neighbors* (reciprocal pairs) introduced by McQuitty (1966). The two points (or groups) a and b are reciprocal neighbors if a is the nearest neighbor of b and b is the nearest neighbor of a. At each step of the basic algorithm, instead of aggregating only the two nearest neighbors, as many new nodes are created as there are reciprocal neighbors. At the final step, all of the points are combined into one group, and the tree is completed.

The problem is then reduced to an efficient search for reciprocal neighbors. We describe the "chain search" algorithm (see Benzecri, 1982).

V.4.1. Algorithm

STEP 1. We begin with an element called x_1. We create a chain of successive elements:

$$x_1, x_2, \ldots, x_{i-1}, x_i, \ldots$$

such that for all i, x_i is a nearest neighbor of x_{i-1}. Such a chain necessarily stops at an element x_k when x_{k-1} is also the nearest neighbor of x_k. Then x_{k-1} and x_k are reciprocal neighbors. They are aggregated to form a node.

STEP 2. If $k = 2$, that is if the chain started with an element that has a reciprocal neighbor, we choose a new element from which a new chain is constructed that stops on new reciprocal neighbors, whose aggregation results in a new node.

STEP 3. If $k > 2$, the search is continued for reciprocal neighbors by extending the chain starting with x_{k-2}. The algorithm stops when $n - 1$ nodes have been created.

V.4.2. Remarks

(a) It has been demonstrated that the maximum cost of the chain search algorithm is αn^2 (where α is a coefficient independent of n), regardless of the configuration of the n points. Juan (1982) developed a FORTRAN program that executes this algorithm.

(b) In order to be able to use the search chain algorithm, the chain must necessarily be extensible beyond x_{k-2} when the reciprocal neighbors x_{k-1} and x_k have been aggregated. It is thus necessary that this aggregation should not destroy the closest neighbor relationship that previously existed between x_{i-1} and x_i $(i = 2, 3, \ldots, k - 2)$. This property is assured if the aggregation law used to construct the tree does not create an *inversion*.

There is no inversion if node $g[a; b]$ created by aggregating a and b cannot be nearer to some element c than are the elements a and b. This condition is written

$$\text{if} \qquad d(a, b) < \inf\{d(a, c), d(b, c)\} \qquad (26)$$

$$\text{then} \quad \inf\{d(a, c), d(b, c)\} < d(g[a; b], c) \qquad (27)$$

This property is verified by several aggregation criteria:

1. Single linkage:

$$d(a, b) = \inf\{d(u, v) | u \in a; v \in b\} \qquad (28)$$

2. Complete linkage:

$$d(a, b) = \sup\{d(u, v) | u \in a; v \in b\} \qquad (29)$$

3. Group average:

$$d(a, b) = \frac{1}{m_a m_b} \sum \{m_u m_v d(u, v) | u \in a, v \in b\} \qquad (30)$$

4. Ward's method:

$$d(a, b) = \frac{m_a m_b}{m_a + m_b} d^2(g_a, g_b) \qquad (31)$$

where g_a and g_b are the centers of gravity of the groups a and b.

V.5.   MIXED METHODS FOR LARGE DATA SETS

The idea of combining the approaches of Sections V.2 (finding partitions) and V.3 (hierarchical trees) is not a new one: it is found, for example, under the name of "hybrid clustering" in Wong (1982). When a large number of objects are to be classified, the following classification strategy is suggested. The first step is to obtain, at a low cost, a partition of the n objects into k homogeneous groups, where k is far greater than the "real" or desired number of groups in the population. The second step is an ascending hierarchical classification, where the terminal elements of the tree are the k groups of the partition. The final step is to visually inspect the tree and cut it at the most appropriate level. The "definitive" partition of the n individuals into k_0 groups ($k_0 < k < n$) is thus obtained.

V.5.1.   Preliminary Partition

The purpose is to obtain rapidly a large number of small groups that are very homogeneous.

We use the partition defined by the stable groups obtained from cross-tabulating a number of base partitions (cf. Section V.2.4). Each base partition is calculated by using the algorithm of moving centers after reading the data directly, so as to minimize the use of central memory. The calculations are generally performed on the coordinates of the individuals on the first few principal axes of an analysis.

V.5.2.   Hierarchical Aggregation of the Stable Groups

Some of the stable groups can be very close to one another, corresponding to a group that is artificially cut by the preceding step. On the other hand, the procedure generally creates several small groups, sometimes containing only one element. The goal of the hierarchical aggregation phase is to reconstitute the groups that have been fragmented, and to aggregate the apparently dispersed elements around their original centers.

The tree is built according to Ward's aggregation criterion (cf. Section V.3.3), which has the advantage of accounting for the size of the elements to classify (as a weight in the calculations of the loss of variance through aggregation).

V.5.3.   Final Partition, and Description of Classes

The final partition of the population is defined by cutting the dendrogram. Choosing the level of the cut, and thus the number of classes in the partition, can be done by looking at the tree: the cut has to be made above

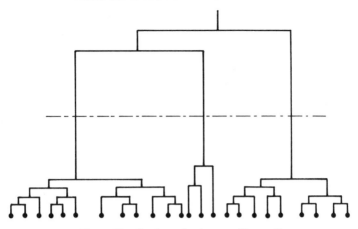

Figure 10.   Cutting a dendrogram "by eye."

the low aggregations, which bring together the elements that are very close to one another, and under the high aggregations, which lump together all of the various groups in the population (Figure 10).

In practice, the situation is often less clearly defined than the one shown in Figure 10 (except when "real," well defined groups exist in the population). The user may vacillate between two or three possible cutting levels.

Each group is then described with the help of the continuous and nominal variables in the original data set. The continuous variables could be ranked by comparing their means within the group to their overall mean. For example, to rank the importance of variable x, one can compare the actual mean with the expected value of the mean, assuming that the individuals were allocated in class j at random (without replacement).

$$t(x) = \frac{\bar{x}_j - \bar{x}}{s_j(x)} \quad \text{with} \quad s_j^2(x) = \frac{n - n_j}{n - 1} \frac{s^2(x)}{n_j} \quad (32)$$

where $s^2(x)$ is the empirical variance of variable x.

The categories of the nominal variables can be ordered in an analogous way. If there are $n_j(m)$ individuals having modality m among the n_j individuals of class j, we may calculate a criterion that is analogous to a t-value and the probability (hypergeometric distribution for N) of having $N \geq n_j(m)$ where the n_j individuals of the group are drawn at random (without replacement) among the population of n individuals.

Table 3a (Section V.6.3) contains sorted values of this criterion for both nominal and continuous variables. The value 11.7 (first row) means that the within-group percentage of the item "age under 25" is approximately 11.7 standard deviations above the global percentage, and consequently highly significant (the value 1.96 of the criterion corresponds to the level 0.05).

V.5.4.   Comments on Classification Strategy

Classifying a large data set is a complex task. It is difficult to find a method (or an algorithm) that alone will surely lead to an optimal result. The proposed strategy, which is not entirely automatic, and which requires several control parameters, allows us to retain control over the classification process.

Similarly, nearest neighbors accelerated algorithms for hierarchical classification permit us to directly build a tree on the entire population. However these algorithms cannot read the data matrix sequentially; the data or the principal coordinates must be stored in central memory. This can be a huge matrix, even for as few as 15 or 20 principal axes.

Besides working with direct reading, the partitioning algorithm has another advantage. The criterion of homogeneity of the groups is better satisfied in finding an optimal partition, rather than in the more constrained case of finding an optimal family of nested partitions (hierarchical tree). In addition, building stable groups constitutes a sort of self-validation of the classification procedure.

V.6.   CLASSIFICATION EXAMPLE

V.6.1.   Data

The data we use here were presented in Chapter IV. 1000 individuals are described by 17 nominal variables and 6 continuous variables. The data were analyzed with multiple correspondence analysis (7 active nominal variables, with a total of 25 categories). We chose the 7 active variables in order to define a "demographic" map of the sample: sex, age, education, and so on.

Our objective is to synthesize the socioeconomic characteristics of the sample by performing a partitioning of the 1000 individuals into several homogeneous groups. The description of the groups constitutes a socioeconomic summary of the data. The advantage of this description is that it goes beyond the two-dimensional representations to which we are restricted by principal axes analysis.

The computations can be done on the raw data, by characterizing each individual by his or her responses to the 7 active variables. In practice, however, there is an advantage in characterizing them by their first coordinates obtained from the preceding multiple correspondence analysis; on the one hand, the last axes generally contain more random noise and peculiarities. On the other hand, distance calculations between individuals are far easier to execute if they are restricted to the first axes of an analysis, when the observations are contained in a higher dimensional space.

V.6.2. Classification Strategy

The classification is performed in several steps. The first step consists of finding a partition of the 1000 individuals that contains a number of groups that is definitely greater than the "real" number of groups. Then the final number of groups is determined by inspecting the hierarchical tree in which these groups are aggregated. We now describe the different phases of the analysis and the parameters needed to perform it.

(a) Aggregation Around Moving Centers. We have chosen to perform the calculations in the space of the first eight factors of the multiple correspondence analysis (the sum of the eight eigenvalues represents 56% of the trace, whereas the sum of the first two factors represents 18% of the trace; see Section VII.3.2 for the interpretation of these percentages).

We perform a first "basic partitioning" of the 1000 individuals into 6 groups around the moving centers (15 iterations were necessary to assure

Table 1.

Three Basic Partitions Into Six Groups	Group Bases After 15 Iterations					
	1	2	3	4	5	6
Partition 1	127	188	229	245	151	60
Partition 2	232	182	213	149	114	110
Partition 3	44	198	325	99	130	204

Table 2.

Stable Groups	Bases (In Decreasing Order)									
1–10	168	118	114	107	88	83	78	26	22	16
11–20	15	14	12	12	12	11	10	7	7	7
21–30	6	6	4	4	4	4	3	3	3	3
31–40	3	3	3	2	2	2	2	2	2	2
41–50	1	1	1	1	1	1	1	1	1	1

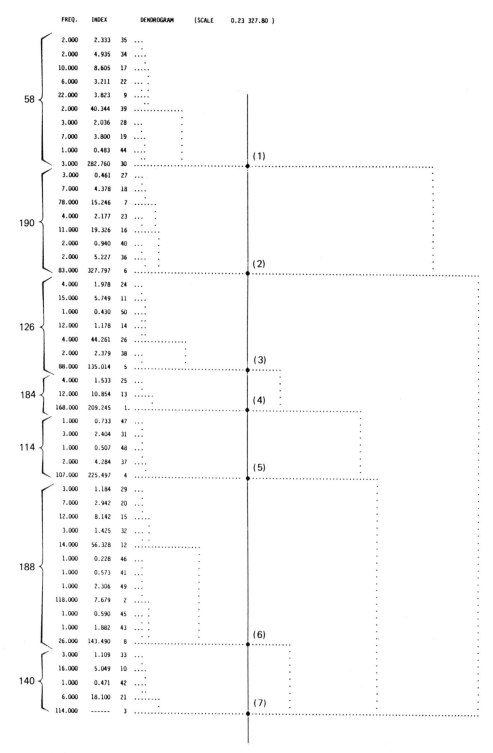

Figure 11. Hierarchical tree of 50 "stable groups."

stability of the groups). This procedure is repeated three times. Table 1 shows the bases of the six groups of the three successive basic partitionings.

Note that the bases tend to fluctuate from one partitioning to the next. This suggests either that the number of "real" groups is different from six, or that the groups are not very distinct within the configuration of points.

(b) Determination of Stable Groups. In reality the preceding step is only an intermediate one. The three partitionings are cross-tabulated, resulting in a partitioning of the 1000 individuals into $6^3 = 216$ groups. The individuals in each of these 216 groups are those who have always been grouped together in the 3 basic partitionings; they constitute the "stable groups." In fact, only 50 groups are not empty, and 40 out of these 50 groups contain less than 15 individuals. The distribution of the individuals is given is Table 2.

(c) Hierarchy for Aggregating the Stable Groups. The 50 groups defined above then undergo a hierarchical classification procedure.

The tree is shown in Figure 11. This tree shows clearly a large number of aggregations at the lowest level, forming groups that in turn become aggregated at a high level, indicating well separated groups.

The final partitioning of the 1000 individuals is obtained by cutting the hierarchical tree at the level at which the distance between groups is judged to be great enough to separate the groups; this cutting is made easier by looking at the histogram of the increasing indices of the aggregations in the tree (Figure 13).

The ideal situation is shown on Figure 12a, where the obvious cut-point is at the threshold of the histogram (this results in five groups because the last bar of the histogram aggregates the two groups at the highest level of

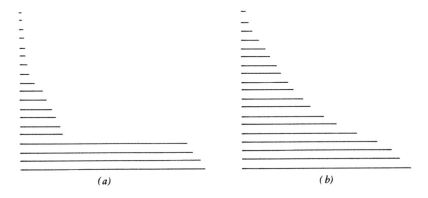

(a) (b)

Figure 12.

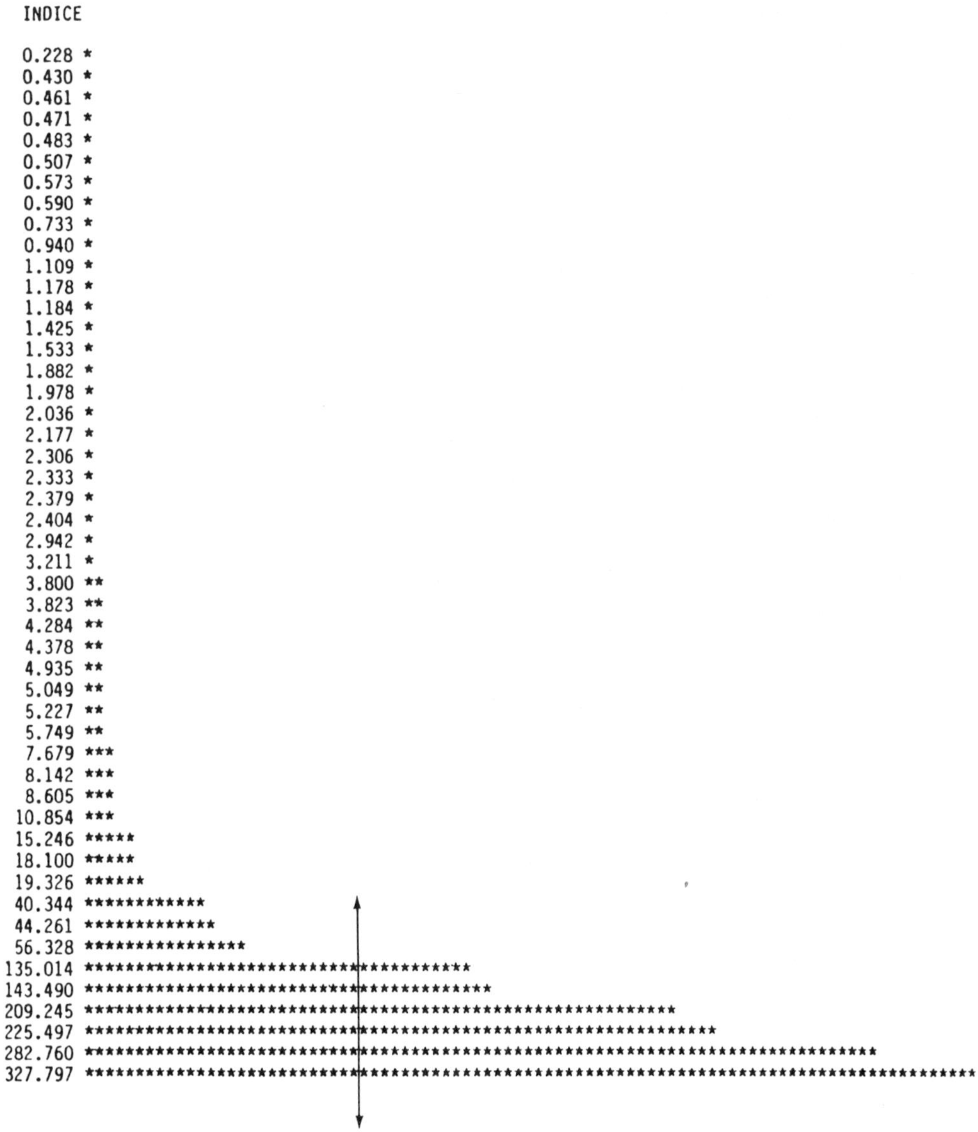

INDICE

```
  0.228 *
  0.430 *
  0.461 *
  0.471 *
  0.483 *
  0.507 *
  0.573 *
  0.590 *
  0.733 *
  0.940 *
  1.109 *
  1.178 *
  1.184 *
  1.425 *
  1.533 *
  1.882 *
  1.978 *
  2.036 *
  2.177 *
  2.306 *
  2.333 *
  2.379 *
  2.404 *
  2.942 *
  3.211 *
  3.800 **
  3.823 **
  4.284 **
  4.378 **
  4.935 **
  5.049 **
  5.227 **
  5.749 **
  7.679 ***
  8.142 ***
  8.605 ***
 10.854 ***
 15.246 ****
 18.100 ****
 19.326 *****
 40.344 ***********
 44.261 ***********
 56.328 **************
135.014 ********************************************
143.490 *******************************************
209.245 ***********************************************************
225.497 **********************************************************
282.760 **********************************************************************************
327.797 ********************************************************************************************
```

Figure 13.   Histogram of indices of hierarchical tree.

the tree). On the other hand, Figure 12b is typical of a situation where it would be impossible to decide on a "real" number of groups in the population.

Between these two criteria, Figure 13 shows a configuration that is rather common (there are one or several thresholds among which the user can choose). Here we have chosen to describe a division into seven groups shown on Figures 11 and 13.

(d) *Note on the Choice of Parameters.* The classification strategy is controlled by the following parameters:

1. The number of basic partitionings, which through cross-tabulation define the stable groups.
2. The number of groups in each basic partitioning.
3. The number of iterations to accomplish the basic partitioning.
4. Finally, the location for cutting the hierarchical tree, in order to determine the number of final groups.

It is difficult to give rules for choosing the values of these parameters. A combination of practical experience and cost considerations suggest choices around the following values:

1. Two or three basic partitionings.
2. A number of groups in each partitioning that approximates the probable number of "real" groups.
3. When the groups are well separated, 6 to 10 iterations are enough to stabilize each basic classification.

V.6.3. Description of the Groups

Table 3a and b contains examples of a computer-generated selection of the items that best characterize groups 1 and 3. This selection, which can be done over hundreds of items (original items, or items developed from cross-classification of original variables), contributes to the heuristic power of the method.

It is very important to keep in mind that the groups are established with a specific set of active elements (here: objective characteristics of the respondents). The supplementary elements are used to identify and to interpret these groups in a second phase.

Let us consider group 1, which contains 58 persons (5.8% of the sample). Although 4% of the population is under 25 years (40 persons), group 1 has over 65% of them. In statistical terms the standard deviation between these proportions (under the hypothesis of homogeneity) is about 12 standard

Table 3a. Description of group 1 sorted on absolute value of criterion (58 observations) (The variables marked with an asterisk are supplementary variables, used after the classification to describe the groups)

Identifiers—Nominal Variables:	Criterion	Frequency	Global	Percentages Of the Category in the Group	Of the Group in the Category
Age: under 25 years	11.7	40	4.0	65.5	95.0
Lodging: free	7.6	58	5.8	43.1	43.1
High school graduate	5.2	162	16.2	44.8	16.0
No stocks and shares	3.5	879	87.9	98.3	6.5
*Ideal number of children: none	2.4	51	5.1	12.1	13.7
*The family is not the only place where we feel good	2.3	431	43.1	56.9	7.7
*Science improves everyday life	2.3	383	38.3	51.7	7.8
*Marriage: breakable by agreement	2.2	387	38.7	51.7	7.8

Identifiers—Continuous Variables	Criterion	Mean Group	Mean Total	Standard Deviation Total
*Age	(−) 8.3	24.2	42.7	17.5
*Number of vacation days	3.0	25.8	18.3	19.4
*Age when leaving school	2.3	18.4	17.3	3.9

Table 3b. Description of group 3 sorted on absolute value of criterion (126 observations) (The variables marked with an asterisk are supplementary variables, used after the classification to describe the groups)

Identifiers—Nominal Variables	Criterion	Frequency	Global	Of the Category in the group	Of the Group in the Category
				Percentages	
Size of town: Less than 2000 inhabitants	14.3	83	8.3	57.1	86.7
Age: 50–64 years	11.1	188	18.8	62.7	42.0
Educational level: grade school	9.4	321	32.1	70.6	27.7
Lodging: owner	8.5	290	29.0	63.5	27.6
*The family is the only place where we feel good	7.1	561	56.1	83.3	18.7
No stocks and shares	4.7	879	87.9	97.6	14.0
*Go to bed before 10 P.M.	4.3	270	27.0	43.7	20.4
*Marriage: unbreakable union	3.5	231	23.1	35.7	19.5

Identifiers—Continuous Variables	Criterion	Mean Group	Mean Total	Standard Deviation Total	
*Age when leaving school	(−) 6.8	15.1	17.3	3.9	
*Age	6.3	51.9	42.7	17.5	
*Number of "no answer" in the survey	4.6	9.4	18.3	19.4	

139

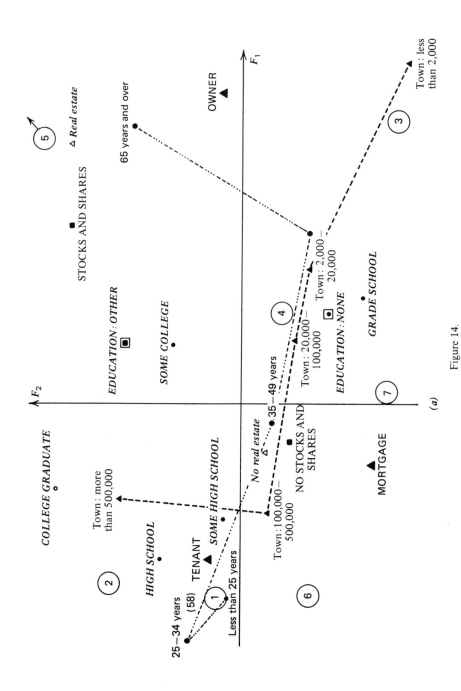

Figure 14.

140

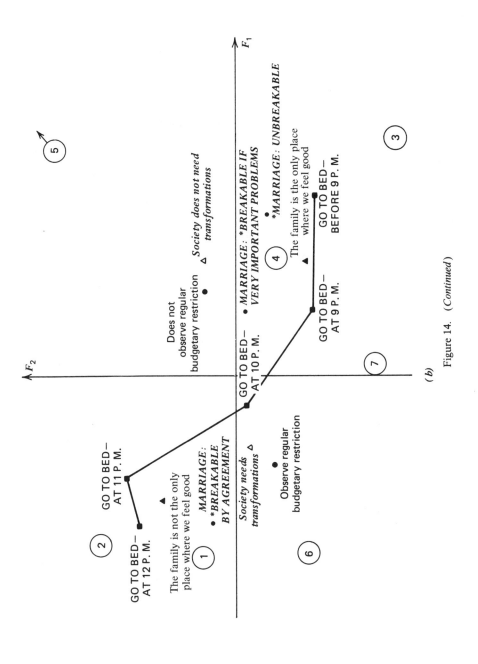

Figure 14. (Continued)

141

deviations. This is shown in the first row of Table 3a. The other characteristics are listed in order of decreasing importance: the group contains significant proportions of persons whose lodgings are free, whose level of education is superior, and so on.

When it is considered as a continuous variable, age is still the primary characteristic of this group. The general mean is approximately 43 years old, and the mean age of the group is slightly over 24 years, thereby deviating from the general mean by over 8 standard deviations (under the null hypothesis of a random selection of group 1 in the population). Similarly, the number of vacation days is greater by over 3 standard deviations from the general mean, and so on.

By contrast, group 3 (Table 3b) appears to be a group of rather elderly persons, living in small towns, often owning their own homes, conservative, and so on. Here is a concise description of the entire typology.

Group 1 (58 persons)	Young people (see Table 3a)
Group 2 (190 persons)	Inhabitants of very large cities, where they are renters; they are high school graduates; they are middle-aged (35–49 years); supplementary elements show that they have given "progressive" responses: the family is not the only place where we feel good, marriage can be broken, society needs changes...
Group 3 (126 persons)	Rather elderly persons, conservatives (see Table 3b)
Group 4 (184 persons)	Similar to group 3, but made up of older people (56% are over 65, as opposed to 17% in the total population)
Group 5 (114 persons)	Older and well-to-do people
Group 6 (188 persons)	Rather young people, renters in medium-size cities, and rather "progressive"
Group 7 (140 persons)	A group of persons who borrowed money to buy their own homes (63% as opposed to 12% in the total population), between 35 and 50 years old, and living in medium-size cities

The link between the classification and the multiple correspondence analysis is made by positioning the centers of the groups in the factorial planes. Figure 14a and b shows the positions of the group centers in the first principal plane (active and illustrative categories). Note the distortion of distances in projection: we see on Figure 11 that group 6, for example, is closer to group 7 (in $\mathbb{R}^8$) than to groups 1 or 2.

Thus the classification and its dendrogram complete the graphical display produced by the principal axes analysis.

V.7.   APPENDIX: EQUIVALENCE BETWEEN SINGLE LINKAGE AND SUBDOMINANT ULTRAMETRIC

To demonstrate this equivalence (cf. Section V.3.2):

1. We define, starting with a distance d, a new distance called the *distance of the smallest maximum jump*.
2. We then show that this distance is an ultrametric distance.
3. Next we show that this ultrametric distance is the *subdominant* one.
4. Finally, we show that this distance corresponds to the ultrametric distance given by the *single linkage* algorithm.

V.7.1.   Definition of the Distance of the Smallest Maximum Jump

Let there be a set E having a distance d. Let x and y be two points contained in E. The pair (x, y) is called the edge, of length $d(x, y)$, of the full graph whose nodes are the elements of E (the name *full graph* comes from the fact that every pair of nodes is joined by an edge). Still using the terminology of graph theory, we call a *path* from x to y a series of edges of the type $(x, t_1), (t_1, t_2), (t_2, t_3), \ldots, (t_{k-1}, t_k), t_k, y)$ where $t_1, \ldots, t_k$ are elements of E. Given a path from x to y, the *maximum jump* is the length of the longest edge of the path from x to y.

To every path joining x and y, there corresponds a maximum jump. Since the set of nodes is finite, there exists a *smallest maximum jump* over the set of paths from x to y; we denote it by $d^*(x, y)$.

V.7.2.   Demonstration that d^* Is an Ultrametric Distance

The smallest maximum jump between x and y is an ultrametric distance. It is clear that the first two axioms of a distance are verified by d^*. To verify that this distance is ultrametric, let us consider three points x, y, and z of E (Figure 15). The smallest maximum jump from x to y, passing through z, is

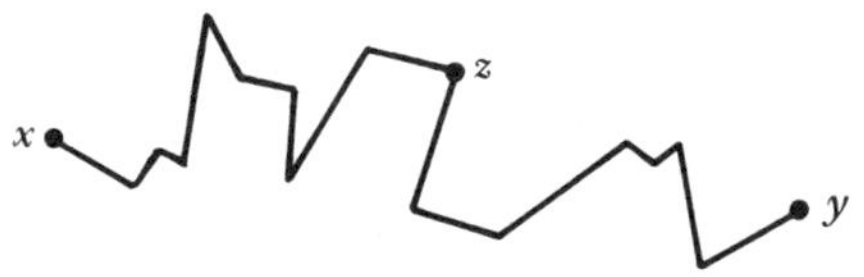

Figure 15.

$\max\{d^*(x, z), d^*(z, y)\}$. The smallest maximum jump from x to y without the constraint of passing through z can only be less than or equal to this quantity; thus

$$d^*(x, y) \leq \max\{d^*(x, z), d^*(y, z)\} \tag{33}$$

and d^* is then shown to be an ultrametric distance.

V.7.3. Demonstration that d^* Is the Subdominant

To show that d^* is the subdominant, we show that d^* is smaller than d and greater than every ultrametric distance that is less than d.

First of all, it is clear that edge (x, y) is a particular path going from x to y; therefore $d^*(x, y) \leq d(x, y)$ and d^* is less than d.

Let d_1 be an ultrametric distance that is smaller than d. For every triplet x_1, x_2, x_3 we obviously have

$$d_1(x_1, x_3) \leq \max\{d_1(x_1, x_2), d_1(x_2, x_3)\} \tag{34}$$

By applying this inequality successively to a path

$$(x_1, x_2), (x_2, x_3), \dots, (x_{p-1}, x_p) \tag{35}$$

we obtain

$$d_1(x_1, x_p) \leq \max_{j<p}\{d_1(x_j, x_{j+1})\} \tag{36}$$

Since $d_1 \leq d$, we have

$$d_1(x_1, x_p) \leq \max_{j<p}\{d(x_j, x_{j+1})\} \tag{37}$$

This inequality is valid for every path joining x_1 to x_p. For at least one of them, we have, by definition of d^*,

$$\max_{j<p}\{d(x_j, x_{j+1})\} = d^*(x_1, x_p) \tag{38}$$

This last equation demonstrates the required inequality.

V.7.4. Equivalence of d^* and d_u

It remains to demonstrate that the ultrametric distance d_u given by the single linkage algorithm is the distance d^* of the smallest maximal jump.

Let $d_u(x, y)$ be the value of the distance at the step where points x and y are joined for the first time. Previously these two points were in different clusters (or constituted clusters in themselves). The method of calculating distances at every aggregation insures that $d_u(x, y)$ is the smallest distance between two points belonging to different clusters.

The distances within a cluster are smaller than $d_u(x, y)$, after aggregation; and the distances involving points that do not belong to the two clusters are greater because they will be aggregated at a future step. The paths joining x and y will therefore have edges inside both clusters, whose length is less than $d_u(x, y)$, and edges outside the clusters whose length is necessarily greater than or equal to $d_u(x, y)$. Thus $d_u(x, y)$ is the smallest maximum jump $d^*(x, y)$.

CHAPTER VI

Direct Reading
Algorithms

This essentially technical chapters deals with *direct reading* algorithms. Such algorithms work by reading a matrix row by row (or column by column) without modifying its structure and without storing large intermediate matrices in the computer's memory.

The advantage of these techniques are twofold:

1. From a *theoretical* viewpoint, they allow results to emerge progressively from the data matrix without any loss of information to the user during the data reduction process.
2. From a *technical* viewpoint, they allow very large data matrices to be analyzed by moderate-size computers.

These algorithms are particularly useful when the coding is *binary coding*: the data matrix is read several times without destroying or modifying it, and the volume of operations is considerably reduced.

We outline briefly the components of this reduced number of computations, particularly with respect to the methods of *clustering around moving centers* (*k*-means) presented in Chapter V. Then we present the techniques of *decomposed iterated power* and *stochastic approximation*.

VI.1. REDUCTION IN THE NUMBER OF OPERATIONS

VI.1.1. General Case

In accordance with the notation in Chapter IV, we shall call $\mathbf{R}$ the (n, Q) matrix containing the n responses to a questionnaire consisting of Q questions, totalling p categories. Thus $\mathbf{R}$ is the *reduced coding* matrix of the (n, p) binary matrix $\mathbf{Z}$.

The basic operation leading to a reduction in the number of operations is the calculation of the *scalar product* between a respondent point of $\mathbf{R}^p$ (i.e., a row of $\mathbf{Z}$) and any vector belonging to this space.

Clearly, this scalar product may be calculated by taking into account only the nonzero terms, of which there are Q: we work only with the Q locations of the 1's in a row of $\mathbf{Z}$, rather than on the p components of this row.

Computation time is then proportional to Q^2 instead of p^2. If each of the questions has 10 categories, this phase of calculations is 100 times faster than using $\mathbf{Z}$. This represents a significant gain if n is large.

The procedures of *decomposed iterated power* and stochastic approximation use essentially the same type of scalar products.

VI.1.2.   The Case of Partitioning Algorithms

In the case of the method of clustering around moving centers, the comparison of the distances of a respondent to the centers of the groups is readily reduced to a comparison of scalar products.

At the mth step, if the center of group k has coordinates C_{kj}^m, the chi-square distance of a respondent i to this center is written (using the notation of Chapter V);

$$d^2(i, k) = nQ \sum_{j=1}^{P} \left(z_{ij}/Q - C_{kj}^m \right)^2 / d_{jj} \qquad \text{where } d_{jj} = \sum_{i=1}^{n} z_{ij} \qquad (1)$$

By expanding the squared term, we find that $d^2(i, k)$ is the sum of there terms:

$$d^2(i, k) = t_2(i) + t_2(k) + t_3(i\,k) \qquad (2)$$

with

$$t_1(i) = \frac{n}{Q} \sum_j \frac{z_{ij}^2}{d_{jj}} \qquad (3)$$

$$t_2(k) = nQ \sum_j \frac{\left(C_{kj}^m \right)^2}{d_{jj}} \qquad (4)$$

$$t_3(i, k) = -2n \sum_j z_{ij} \frac{C_{kj}^m}{d_{jj}} \qquad (5)$$

For each respondent i, we have to find the smallest $d^2(i, k)$ over the k_c groups ($k \leq k_c$).

We see that, in this comparison, the term $t_1(i)$ is irrelevant. The term $t_2(k)$ is independent of i, and thus can be calculated once and for all at the beginning of iteration m. The only term that has to be calculated for each respondent and for each group is $t_3(i, k)$, which is a scalar product of the type mentioned above: it can be evaluated in Q operations instead of p operations.

Thus the cost of a partition obtained by these methods depends only on the number Q of questions, independently of the number of possible responses to these questions.

VI.2. SIMULTANEOUS ITERATED POWER ALGORITHM (HOTELLING, 1933)

Hotelling's algorithm for finding the eigenvalues and eigenvectors of a symmetric matrix, called the iterated power algorithm, has a relatively slow rate of convergence. It is practically never used any more when the matrix to be diagonalized can fit into the computer's memory: it has been supplanted by Jacobi's method, and by techniques that imply a preliminary tridiagonalization of the matrix, the principle of which originated with Householder (cf. Wilkinson, 1965).

Yet this method has the advantage of not being a "black box," because it is possible to interpret its process. Additionally, it works in direct reading mode, if the matrix to be diagonalized is calculated anew with every iteration.

Let us briefly review the features of this algorithm.

Let A be the (p, p) nonnegative square matrix to be diagonalized. If $X_0 = (x_0^1, x_0^2, \ldots, x_0^r)$ is a basis of a subspace of $\mathbb{R}^p$, the algorithm consists of constructing the series

$$X_1 = AX_0 \tag{6}$$

$$X_2 = AX_1^* \tag{7}$$

$$\vdots$$

$$X_k = AX_{k-1}^* \tag{8}$$

where X_j^* is obtained from X_j by an orthonormalization algorithm.

In fact, the efficiency of the method is greatly increased if the iterated power is used simply to provide an approximation of the subspace $\mathscr{V}_q$ with q

dimensions containing the first r principal axes. It is no longer necessary to orthonormalize at every step; on the other hand, the matrix $X_k' A X_k$, whose size is (q, q), still must be diagonalized at iteration k.

Let us note λ_α, the αth eigenvalue of A.

The rate of convergence will be in fact governed by $|\lambda_{q+1}/\lambda_r|$. This ratio can be much smaller if we introduce a suitable shift of origin (c.f. Wilkinson, 1965). This ratio becomes $|(\lambda_{q+1} - \beta)/(\lambda_r - \beta)|$. (The transformation is obtained by replacing X_k by $X_k - \beta X_{k-1}$ at the end of each iteration.) Without a priori knowledge, a commonly adopted value is $\beta = $ (trace $A)/2p$ (half the average eigenvalue). Since we do not intend to directly estimate the eigenvalues, it is not necessary to orthonormalize X_k after each iteration. However this inexpensive operation is recommended to reduce the effects of rounding errors.

VI.2.1. First Decomposition of Matrix A (Reciprocal Averaging)

In correspondence analysis matrix A is the product of two *probabilistic transitions*, that is, the product of two matrices of conditional frequencies.

According to the results of Section II.2.2, the factor φ to be found is an eigenvector of

$$A = \left(D_p^{-1}F'\right)\left(D_n^{-1}F\right) \tag{9}$$

Starting with any eigenvector φ_0, the algorithm consists of calculating successively

$$\psi_0 = \left(D_n^{-1}F\right)\varphi_0 \tag{10}$$

$$\varphi_1 = \left(D_p^{-1}F'\right)\psi_0 \tag{11}$$

$$\psi_1 = \left(D_n^{-1}F\right)\varphi_1 \tag{12}$$

$$\vdots$$

With this method, the barycenters relative to the two sets are calculated alternately. Thus the presentation of correspondence analysis in Section II.2.5 suggests a computational algorithm that is particularly simple but not very efficient (Richardson and Kuder, 1933; Hill, 1973, 1974).

VI.2.2. Second Decomposition of Matrix A

We limit ourselves to the case where an (n, p) matrix Z, with binary coding, is analyzed. Matrix A is the sum of n matrices of rank 1.

If z_i' designates the ith row (with p components) of the data matrix $\mathbf{Z}$, and if $\mathbf{D}$ is the diagonal matrix whose diagonal elements are the sums of the columns of $\mathbf{Z}$, the following is true in the correspondence analysis of matrix $\mathbf{Z}$ (see Section IV.2.1):

$$\mathbf{A} = \frac{1}{Q}\mathbf{D}^{-1}\mathbf{Z}'\mathbf{Z} = \frac{1}{Q}\mathbf{D}^{-1}\sum_{i=1}^{n}\mathbf{z}_i\mathbf{z}_i' \tag{13}$$

The equation

$$\boldsymbol{\varphi}_1 = \mathbf{A}\boldsymbol{\varphi}_0 \tag{14}$$

is written

$$\boldsymbol{\varphi}_1 = \frac{1}{Q}\sum_{i=1}^{n}\mathbf{D}^{-1}\mathbf{z}_i\left(\mathbf{z}_i'\boldsymbol{\varphi}_0\right) \tag{15}$$

Let us consider point i of $\mathbb{R}^p$, whose coordinates are the components of $(1/Q)\mathbf{z}_i$. The scalar quantity

$$s(i, \boldsymbol{\varphi}_0) = \frac{1}{Q}\mathbf{z}_i'\boldsymbol{\varphi}_0 \tag{16}$$

is the abscissa of the projection of this point i on the axis corresponding to the factor $\boldsymbol{\varphi}_0$.

This scalar product can be calculated in Q operations, instead of p. The vector

$$\mathbf{p}_i = n\mathbf{D}^{-1}\mathbf{z}_i \tag{17}$$

is associated with the projection operator on the straight line connecting the origin to point i. Therefore we have

$$\boldsymbol{\varphi}_1 = \frac{1}{n}\sum_{i=1}^{n}s(i, \boldsymbol{\varphi}_0)\mathbf{p}_i \tag{18}$$

Thus $\boldsymbol{\varphi}_1$ is a weighted mean of the operators $\mathbf{p}_i$. In practice this mean is calculated by performing n times, during iteration k, an assignment such that

$$\boldsymbol{\varphi}_k \leftarrow \boldsymbol{\varphi}_k + \frac{1}{n}s(i, \boldsymbol{\varphi}_{k-1})\mathbf{p}_i \tag{19}$$

Vector $\boldsymbol{\varphi}_{k-1}$ remains unchanged during iteration k.

Note that the information gathered during the row by row reading of matrix Z is not taken into account in a sequential manner.

VI.3.   STOCHASTIC APPROXIMATION

The algorithm for diagonalization by stochastic approximation was first presented by Benzecri (1969b) in a general probabilistic framework. However, it is a distant relative of the nonparametric iterative estimation techniques that were proposed by Robbins and Monro (1951).

The general principle of the algorithm is as follows: equation (19) above, which describes the kth row by row reading of matrix Z is substituted for equation (20):

$$\varphi_{k,\,i+1} \leftarrow \varphi_{k,\,i} + h(i,k)s(i,\varphi_{k,\,i})\mathbf{p}_i \qquad (20)$$

where $h(i,k)$ is a decreasing function of i, and a strictly decreasing function of k; its form is generally

$$h(i,k) = \frac{c}{(i+(k-1)n)^{\gamma}} \qquad \text{with } c > 1 \text{ and } 0.5 < \gamma \le 1 \qquad (21)$$

Thus *estimation of factor φ is modified after each individual is read*, and only takes into account the value that is determined at the time the preceding individual is read.

We can say that the simple reading of the iterated power [equation (19)], followed by an overall computation, is substituted by a Markovian read. This type of reading is probably closer to the psychological process of a reader eyeballing a table in order to glean its essential points.

The difference between iterated power and stochastic approximation is of the same order as the difference between the two methods of clustering around moving centers, Forgy (1965) method, versus the k-mean method as developed particularly by MacQueen (1967). In the first type of method, the moving centers remain unchanged during the row by row reading of a matrix. In the second type, these centers are recomputed after each individual is read. However, the fact that these methods both converge toward local optima makes it extremely difficult to compare them in terms of efficiency.

VI.3.1.   Definitions and Notation

Let $\mathbf{x}$ be a vector of $\mathbb{R}^p$, and $\mathbf{A}$ a (p, p) matrix.

Corresponding to any vector norm we can define a nonnegative quantity

$$\frac{\underset{\mathbf{x}\neq 0}{\sup} \|\mathbf{Ax}\|}{\|\mathbf{x}\|} \qquad (22)$$

This is equivalent to

$$\sup_{\|\mathbf{x}\|=1} \|\mathbf{Ax}\| \tag{23}$$

(vid. Wilkinson, 1965). This is termed the matrix norm subordinate to the vector norm.

$$\|\mathbf{A}\| = \frac{\sup\limits_{\mathbf{x} \neq 0} \|\mathbf{Ax}\|}{\|\mathbf{x}\|} \tag{24}$$

Thus

$$\|\mathbf{Ax}\| \leq \|\mathbf{A}\| \cdot \|\mathbf{x}\| \tag{25}$$

for all nonzero $\mathbf{x}$. Matrix and vector norms for which equation (25) is true are said to be compatible. Therefore a vector norm and its subordinate matrix norm are always compatible.

We only consider a finite string of the n elements $\mathbf{A}(i)$. We note

$$\mathbf{A} = \sum_{i=1}^{n} p(i)\mathbf{A}(i) \tag{26}$$

where $p(i)$ is the relative weight of observation i. (in the case of binary coding, and in what follows, $p(i) = 1/n$ for all i).

We call iteration k the set of operations performed during the kth read of the string $\{\mathbf{A}(i), i = 1, \ldots, n\}$.

The identity matrix is called $\mathbf{I}$.

VI.3.2. Convergence of the Algorithm

The stochastic approximation algorithm involves the function $h(i, k)$ defined above. However, the algorithm works satisfactorily if $h(i, k)$ is constant during an iteration; in this case, $h(i, k) = h(k)$.

We consider the case of finding one factor φ. After iteration $k - 1$, the estimate for this factor φ is φ_{k-1}. Iteration k consists of premultiplying the vector φ_{k-1} by operator $\mathbf{H}(k)$, which is defined as follows:

$$\mathbf{H}(k) = \prod_{i=1}^{n} (\mathbf{I} + h(i, k)\mathbf{A}(i)) \tag{27}$$

Even though the framework of this discussion is numerical, as opposed to probabilistic, the demonstration of convergence is analogous to that found in Benzecri (1969b). In particular, *lemma 0*, which is established by this author, remains unchanged. We do not give its proof here.

Lemma 0. Given the definition of the norm given above, if $\|\mathbf{V}_k\| > 1$, and if $\sum \|\mathbf{v}_k\| < 1$, we have the inequality

$$\left\| \prod_k \mathbf{V}_k - \prod_k (\mathbf{V}_k + \mathbf{v}_k) \right\| < \prod_k \|\mathbf{V}_k\| 2 \sum_k \|\mathbf{v}_k\| \qquad (28)$$

First we consider the case of a function $h(i, k)$, which is constant during an iteration (i.e., during a reading of the entire matrix row by row).

(a) Case of a Function $h(i, k) = h(k)$

Lemma 1. If the elements $\mathbf{A}(i)$ are such that: for $i \leq n$, $\|\mathbf{A}(i)\| < a$, and m designates a real number greater than or equal to na; if in addition $\|\mathbf{A}\| \leq 1$, then

$$\left\| \prod_{i=1}^{n} \left(\mathbf{I} + \frac{\mathbf{A}(i)}{m} \right) - \exp\left\{ \frac{n}{m} \mathbf{A} \right\} \right\| < (a^2 + 1)\left(\frac{n}{m} \right)^2 \qquad (29)$$

The proof of this lemma is in Section VI.5.1 (the Appendix to this chapter).

To use the inequality of lemma 0, suppose

$$\mathbf{V}_k = \exp\{ nh(k)\mathbf{A} \} \qquad (30)$$

$$\mathbf{V}_k + \mathbf{v}_k = \mathbf{H}(k) = \prod_{i=1}^{n} (\mathbf{I} + h(k)\mathbf{A}(i)) \qquad (31)$$

Lemma 1 then gives us an upper bound for $\|\mathbf{v}_k\|$, if we suppose that $1/m = h(k)$:

$$\|\mathbf{v}_k\| < (a^2 + 1)[h(k)n]^2 \qquad (32)$$

The quantity k_1 is chosen to ensure that

$$\sum_{k_1}^{k_2} \|\mathbf{v}_k\| < 1 \qquad (33)$$

This is always possible because the series $h(k)$ is chosen so as to ensure the convergence of the series $\Sigma\|\mathbf{v}_k\|$. Then equation (28) is written

$$\left\|\prod_{k=k_1}^{k=k_2} \mathbf{H}(k) - \exp\left\{\sum_{k_1}^{k_2} h(k)n\mathbf{A}\right\}\right\| \leq 2(a^2 + 1)n^2\exp\left\{\sum_{k_1}^{k_2} h(k)n\right\}\sum_{k_1}^{k_2} h^2(k) \tag{34}$$

Let us construct a series of elements of $\mathbb{R}^p$ with a general term φ_j starting with any vector φ_0 in $\mathbb{R}^p$, with the formula

$$\varphi_j = \varphi_{j-1} + h(k)\mathbf{A}(i)\varphi_{j-1} \tag{35}$$

where the indices are such that

$$j = i + (k-1)n \tag{36}$$

Index i varies between 1 and n, and index k increases by 1 every time i takes the value 1 after reaching the value n.

Proposition 1. If the conditions of Lemma 1 are satisfied, and if φ_j designates the series defined above, vector φ_j tends toward a vector that is collinear with the first eigenvector of $\mathbf{A}$, when j increases indefinitely, if the series of general term $h(k)$ diverges and the series of general term $h^2(k)$ converges. Note that we are assuming that the largest eigenvalue of $\mathbf{A}$ is unique.

Let us briefly sketch the proof.

If the conditions of Proposition 1 are satisfied, we can find an integer K such that, for all k_1 and k_2 greater than K, the term $\Sigma_{k_1}^{k_2}h^2(k)$ of the right side of equation (34) is as small as we want it to be, and that the quantity

$$\sum_{k_1}^{k_2} h(k) \tag{37}$$

is as large as we want it to be.

The operator

$$\exp\left\{\sum_{k_1}^{k_2} h(k)n\mathbf{A}\right\} \tag{38}$$

must be normalized (because its largest eigenvalue increases indefinitely). The scale factor is the quantity

$$\exp\left\{\sum_{k_1}^{k_2} h(k)n\right\} \tag{39}$$

which appears in the right-hand side of equation (34). After normalization, this operator tends toward a mapping of $\mathbb{R}^p$ onto the first eigenvector of $\mathbf{A}$, while remaining arbitrarily near the operator (which is normalized the same way):

$$\prod_{k_1}^{k_2} \mathbf{H}(k) \tag{40}$$

Example (suggested by Benzecri). Assume

$$h(k) = \frac{1}{n} h'(k) \tag{41}$$

with

$$h'(k) = 1 + \left(\tfrac{1}{2} + \tfrac{1}{2}\right) + \left(\tfrac{1}{4} + \tfrac{1}{4} + \tfrac{1}{4} + \tfrac{1}{4}\right) + \left(\tfrac{1}{8} + \cdots + \tfrac{1}{8}\right) + \cdots \tag{42}$$

From the rank $k_1 = 2^{r-1}$ to the rank $k_1' = 2^r - 1$ (including these two ranks), the general term of this series is $\left(\tfrac{1}{2}\right)^{r-1}$. Assume

$$k_1 = 2^{r-1} \quad \text{and} \quad k_2 = 2^{p-1} \tag{43}$$

with $p > r$. Under these conditions, the sum

$$\sum_{k_1}^{k_2} h'(k) \tag{44}$$

is equal to $p - r$ (since the sum of these terms in each set of parentheses is equal to 1). The series $h'^2(k)$, after rearranging the terms inside the parentheses, is the geometric progression with a power $\tfrac{1}{2}$.

The sum

$$\sum_{k_1}^{k_2} h'^2(k) \tag{45}$$

is equal to

$$\tfrac{1}{2}^{r-1}\left(1 - \left(\tfrac{1}{2}\right)^{p-r+1}\right) < \frac{1}{k_1} \tag{46}$$

which leads to the explicit inequality

$$\left\| \prod_{k_1}^{k_2} \mathbf{H}(k) - \exp\left\{\log_2 \frac{k_2}{k_1} \mathbf{A}\right\} \right\| < \frac{2}{k_1}(a + 1)\exp\left\{\log_2 \frac{k_2}{k_1}\right\} \tag{47}$$

(b)   Case of a General Function h(i, k).   In the case where the function h is not constant during an iteration, the inequality of Lemma 1 has to be modified, since the first degree terms do not cancel each other out identically on the left side of equation (29).

We examine the previous case where

$$h(i, k) = \frac{c}{(i + (k - 1)n)^\gamma} \tag{48}$$

Suppose

$$m = (k - 1)n$$

$$T(m, \gamma, c) = \sum_{i=1}^{n} \frac{c}{(m + i)^\gamma} \tag{49}$$

Lemma 2.   If the series $A(i)$ is such that $\|A(i)\| \le a$ for all $i \le n$; and if m designates a real number such that $m^\gamma > can$; we have, by supposing that $\|A\| \le 1$;

$$\left\| \prod_{i=1}^{n} \left(I + \frac{cA(i)}{(m + i)^\gamma} \right) - \exp\{ T(m\,\gamma, c)A \} \right\|$$

$$\le c^2(a^2 + 1)\left(\frac{n}{m^\gamma} \right)^2 + \frac{2a\gamma cn^2}{m(m + n)^\gamma} \tag{51}$$

The proof of this lemma is very similar to that of Lemma 1.

The process converges for reasons given in Proposition 1, if $\gamma < 1$, so as to ensure the divergence of the series $\Sigma\{ T(n(k - 1), \gamma, c)|k \le K \}$, and if $\gamma > \frac{1}{2}$, so as to ensure the convergence of the series of the upper bounds. The fact that the series $h(i, k)$ varies within an iteration does not modify the conditions of convergence of the algorithm.

(c)   Other Properties of the Algorithm.   In Section VI.5, the Technical Appendix, it is shown that the quality of the convergence can be improved by reading the file in different directions, but this property is not used in the programs. The Appendix also shows the possibility of obtaining several eigenvectors simultaneously.

VI.3.3.   Comparison with Iterated Power

We have seen that iteration k consists in subjecting vector φ_{k-1} (whose norm is supposed to be 1), resulting from previous iterations, to the operator

$$H(k) = \prod_{i=1}^{n} (I + h(i, k)A(i)) \tag{52}$$

If φ designates the required eigenvector (of norm 1), this operation, when compared with iterated power, is only advantageous if

$$\cos(\varphi, \mathbf{H}(k)\varphi_{k-1}) > \cos(\varphi, \mathbf{A}\varphi_{k-1}) \tag{53}$$

If we assume that $\|\mathbf{A}\| = 1$, a simple calculation shows that $\mathbf{H}(k)$ must no longer be used when $\alpha(k) < 1$ with

$$\alpha(k) = \sum_{i=1}^{n} \frac{c}{(i + n(k - 1))^{\gamma}} \tag{54}$$

(a)   For example, assume that $c = 1$, $n = 10^3$, $\gamma = 1$ (therefore $h(i, k)$ is the harmonic series). For the first iteration

$$\alpha \approx \log 10^3 + \mathscr{C} \gg 1 \qquad (\mathscr{C} = \text{Euler's constant}) \tag{55}$$

For the second iteration;

$$\alpha \approx \log 2 < 1 \tag{56}$$

Thus the stochastic approximation must be dropped at the second iteration.

(b)   When $\gamma = 1$, and in general, for $k > 1$;

$$\alpha(k) \approx c \log\left(\frac{k}{k - 1}\right) < \frac{c}{k - 1} \tag{57}$$

It is therefore necessary (but not sufficient) that constant c be greater than $k - 1$ for the kth iteration to be more advantageous than the iterated power.

(c)   If $\gamma < 1$, we have

$$\alpha(k) \approx \frac{c}{1 - \gamma} n^{1-\gamma}\left[k^{1-\gamma} - (k - 1)^{1-\gamma}\right] < \frac{cn^{1-\gamma}}{k^{\gamma}} \tag{58}$$

A numerical example is given in Section VI.5.

VI.4.   PERFORMING THE CALCULATIONS

As shown in Section VI.3.3, the asymptotic efficiency of the stochastic approximation is weak. In practice, we use the stochastic approximation algorithm only as a device for accelerating the iterated power algorithm used simultaneously.

These two procedures are not used to calculate the eigenvectors and eigenvalues, but to find a q dimensional subspace, that contains the r first eigenvectors ($r < q$). This subspace $\mathscr{V}_q$ is characterized by a basis whose vectors are the q columns of a matrix **B**. The different steps of the procedure can be summarized as follows ($v(\mathbf{B})$ is the same as $\mathscr{V}_q$):

1. First estimate of the (p, q) matrix **B** through stochastic approximation. Let us note $\mathbf{B}_1$ this estimate.
2. Improvement of $\mathbf{B}_1$ through k simultaneous iterations, with shift of origin (producing $\mathbf{B}_2$).
3. First estimate of the r dominant eigenvectors and eigenvalues after projection onto $v(\mathbf{B}_2)$ (diagonalization using a classical in-core algorithm).
4. Performance of a supplementary simultaneous iteration.
5. Second estimate of the r eigenvectors, and computation of the cosines of their angles with the previous estimates.

Two parameters are used to control the process: q, dimension of the subspace $v(\mathbf{B})$, and k, the number of iterations during step 2. Usual values giving satisfactory results are $q = r + 5$ and $k = 10$.

VI.5.   TECHNICAL APPENDIX

VI.5.1.   Proof of Lemma 1

The equation

$$nA = \sum_{i=1}^{n} A(i) \tag{59}$$

allows us to find an upper bound of the left side with the quantity

$$\prod_{i=1}^{n}\left(1 + \left\|\frac{A(i)}{m}\right\|\right) - 1 - \sum_{i=1}^{n}\left\|\frac{A(i)}{m}\right\| + \exp\left\{\frac{n}{m}\|A\|\right\} - 1 - \frac{n}{m}\|A\| \tag{60}$$

and this expression is less than

$$\left(1 + \frac{a}{m}\right)^n - 1 - \frac{an}{m} + \exp\left\{\frac{n}{m}\right\} - 1 - \frac{n}{m} \tag{61}$$

However:

$$\left(1 + \frac{a}{m}\right)^n \le \exp\left\{\frac{an}{m}\right\} \le 1 + \frac{an}{m} + \left(\frac{an}{m}\right)^2 \tag{62}$$

from which we deduce the inequality of Lemma 1.

VI.5.2. Discussion of Back-and-Forth Readings

One of the main obstacles in finding inequalities is the noncommutativity of the algebra of matrices. It is interesting to note that by alternating the direction of reading the matrix, the inequalities are improved, by achieving some of the simplifications that commutativity would allow.

We study this procedure in the case where the function h is constant during a pair of iterations, which makes it easier to calculate the inequalities and simplifies the notation.

Lemma 3. If the elements of the finite series $A(i)$ always satisfy $\|A(i)\| \le a$, for all $i \le n$, and $\|A\| \le 1$; if m designates a real number such that $m \ge 2an$, we have the inequality

$$\left\| \prod_{i=1}^{n} \left(I + \frac{A(i)}{m} \right) \prod_{i=n}^{1} \left(I + \frac{A(i)}{m} \right) - \exp\left\{ \frac{2n}{m} A \right\} \right\| \le \frac{a^2 n}{m^2} + \frac{a^3 + 1}{3} \left(\frac{2n}{m} \right)^3$$

$$(63)$$

For the proof of this lemma, see Lebart (1974).

Example. If at the kth iteration $m = 2kn$, and if $a = 1$, then the upper bound is written

$$\frac{1}{4k^2 n} + \frac{2}{3k^3} \qquad (64)$$

From a practical point of view, the second degree term in $1/k$ long remains insignificant with respect to the third degree term, since n is often very large in actual applications.

Note that, if the directions had not been reversed, the inequality given by lemma 1 would be

$$(a^2 + 1)\left(\frac{2n}{m} \right)^2 \qquad \text{or} \qquad \frac{2}{k^2} \qquad (65)$$

VI.5.3. Simultaneous Determination of q Principal Axes

In practice, $A(i)$ is a matrix of rank 1, of the type $z_i z_i'$ and A is positive definite. We consider this case, which leads to some simplifications.

Let us designate by

$$q\Lambda\left(I + \frac{A(i)}{m} \right) \qquad (66)$$

the qth exterior power of the operator $(\mathbf{I} + \mathbf{A}(i)/m)$. This exterior power is defined by the relationship

$$^q\Lambda\left(\mathbf{I} + \frac{\mathbf{A}(i)}{m}\right)(\mathbf{x}_1\Lambda\mathbf{x}_2\Lambda\ldots\Lambda\mathbf{x}_q) = \left(\mathbf{I} + \frac{\mathbf{A}(i)}{m}\right)\mathbf{x}_1\Lambda\ldots\Lambda\left(\mathbf{I} + \frac{\mathbf{A}(i)}{m}\right)\mathbf{x}_q \tag{67}$$

Since $\mathbf{A}(i)$ has a rank of 1, we have the identity between endomorphisms of $^q\Lambda E$ (the unit operator is designated by $\mathbf{I}$ in E as well as in $^q\Lambda E$).

$$^q\Lambda\left(\mathbf{I} + \frac{\mathbf{A}(i)}{m}\right) = \mathbf{I} + \frac{[\mathbf{A}(i)]}{m} \tag{68}$$

In this expression $[\mathbf{A}(i)]$ is defined by

$$[\mathbf{A}(i)](\mathbf{x}_1\Lambda\mathbf{x}_2\Lambda\ldots\Lambda\mathbf{x}_q) = \sum_{h=1}^{q} \mathbf{x}_1\Lambda\mathbf{x}_2\ldots\Lambda\mathbf{A}(i)\mathbf{x}_h\Lambda\ldots\Lambda\mathbf{x}_q \tag{69}$$

for every element of $^q\Lambda E$. With this notation, we have

$$[\mathbf{A}] = \frac{1}{n} \sum_{i=1}^{n} [\mathbf{A}(i)] \tag{70}$$

The conditions of Lemma 1 lead to the following inequalities when $^q\Lambda E$ is provided with a norm analogous to that of E

$$\|[\mathbf{A}(i)]\| < qa \tag{71}$$

$$\|[\mathbf{A}]\| < q \tag{72}$$

Lemma 1 then takes the following form.

Lemma 4.   If the elements $\mathbf{A}(i)$ are such that, for all i, $\|\mathbf{A}(i)\| < a$; $\|\mathbf{A}\| < 1$; and if m designates a real number such that $m \geq qa$ where q is an integer less than the dimension p of E, we have (using the previous notation)

$$\left\| \prod_{i=1}^{n} \left\{ {}^q\Lambda\left(\mathbf{I} + \frac{\mathbf{A}(i)}{m}\right)\right\} - \exp\left\{\frac{n}{m}[\mathbf{A}]\right\} \right\| \leq q^2(a^2 + 1)\left(\frac{n}{m}\right)^2 \tag{73}$$

The eigenvectors of $[\mathbf{A}]$ are the same as those of $^q\Lambda\mathbf{A}$: they are the exterior products of q-tuples of eigenvectors of $\mathbf{A}$; to the vector

$$\mathbf{x}_{\alpha 1}\Lambda\mathbf{x}_{\alpha 2}\Lambda\ldots\Lambda\mathbf{x}_{\alpha_q} \qquad (\alpha_1 < \alpha_2 < \cdots < \alpha_q) \tag{74}$$

corresponds the eigenvalue:

$$\lambda_{\alpha_1}\lambda_{\alpha_2}\ldots\lambda_{\alpha_q} \qquad \text{for } {}^q\Lambda\mathbf{A} \tag{75}$$

and the eigenvalue

$$\lambda_{\alpha_1} + \lambda_{\alpha_2} + \cdots + \lambda_{\alpha_q} \quad \text{for } [\mathbf{A}] \tag{76}$$

The first eigenvector of $[\mathbf{A}]$, and therefore of $\exp\{(n/m)[\mathbf{A}]\}$, consequently entirely characterizes the subspace generated by the q first eigenvectors of $\mathbf{A}$, provided that all the eigenvalues of $\mathbf{A}$ are distinct.

Thus the demonstration of the convergence in ${}^q\Lambda E$ is in every way analogous to the one already done in E, by substituting $\mathbf{A}(i)$ with $[\mathbf{A}(i)]$ slightly modifying the inequalities according to the inequalities of Lemma 4.

This allows us to obtain, starting with a q-tuple of vectors of E, an approximation of the q-dimensional space generated by the q principal axes.

VI.5.4. Numerical Example

The test data involves a data matrix with the following characteristics: $Q = 5$; $p = 43$; $n = 1000$.

The first two eigenvalues are $\lambda_1 = 0.453716$ and $\lambda_2 = 0.411627$.

We compare the simultaneous iterated power, the stochastic approximation with a simple reading, the back-and-forth reading, and the combined techniques. The cosines of the angles between the real axes and the estimated axes are measured at iteration k (cf. Table 1). The results are numerically acceptable. We see that the first stochastic approximation iteration leads directly to the level of the fifth iterated power reading.

By alternating the direction of the reading the level of the sixth simple reading is reached in three readings.

Finally, if the stochastic approximation (whose asymptotic efficiency is weak) is replaced by the iterated power starting with the third reading, we obtain, after a total of six readings, cosines of 0.999997 and 0.999996, and estimates of the first two eigenvalues of 0.453716 and 0.411626 (or a difference of less than 10^{-6} with the actual values).

Table 1. Cosines between real axes and calculated axes

Iteration (Reading)	Iterated Power		Stochastic Approximation (Simple Reading)		Stochastic Approximation (Alternating Reading)	
	f_1	f_2	f_1	f_2	f_1	f_2
1	0.59	0.42	0.992	0.991	0.992	0.991
2	0.84	0.72	0.998	0.997	0.9995	0.9998
3	0.95	0.90	0.999	0.998	0.9998	0.9998
4	0.990	0.96	0.9993	0.9992	0.99994	0.99996
5	0.997	0.990	0.9996	0.9994	0.99996	0.99996
6	0.9990	0.995	0.9997	0.9996	0.99998	0.99998

CHAPTER VII

Reliability and Significance of Results

In this chapter we attempt to answer the following questions:

1. Which data matrices should we analyze? How do we construct these matrices?
2. What can we expect from multivariate descriptive statistical analysis (MDSA) techniques?
3. How do we evaluate the quality of the configurations we obtain?

We limit ourselves to the methods presented in the preceding chapters, that is, to the techniques of descriptive principal components analysis, correspondence analysis and its extensions, and partitioning techniques. The reason for this choice is that they are inexpensive and readily interpreted and thus are the most fruitful techniques for describing large matrices.

VII.1. WHICH DATA MATRICES SHOULD WE ANALYZE? HOW DO WE CONSTRUCT THEM?

These methods are most appropriate in the following situation: we want to describe large, homogeneous matrices (of measurements, ratings, or codes) about which very little is known a priori.

Three conditions should exist:

1. The matrix must be so *large* that visual inspection or elementary statistical analyses cannot reveal its structure.
2. It must be *homogeneous*, so that it is appropriate to calculate statistical distances between its rows and its columns, and so that these distances can be meaningfully interpreted.

162

3. It must be *amorphous*, a priori; this means that applying these methods is most useful where the structure of the matrix is unknown or only partially understood.

The property of homogeneity needs further elaboration. It is usually understood as homogeneity of the *texture* of the matrix; the coding of the matrix must allow the rows or columns to be comparable; for example, quantities expressed in grams and in meters should not be mixed.

Textural homogeneity can generally be achieved through analytical transformations or appropriate coding. Thus *normed* principal components analysis allows us to analyze heterogeneous measurements (with disparate scales) by standardizing the original variables (cf. Section I.3.1).

Transformation into ranks (provided the initial variables have a more or less continuous distribution, with few ties) allows us to increase, even more, the homogeneity of the matrix to be analyzed (cf. Section I.3.4).

Binary coding (cf. Chapter IV) allows correspondence analysis to simultaneously analyze nominal variables (such as region or socioprofessional category) and continuous variables (age, income) that have previously been coded into classes.

However, textural homogeneity is generally not sufficient. For a clear interpretation, it is important that the material being analyzed should be homogeneous in its *substance*, or, rather, its *content*, thus respecting the *principle of relevance* recommended by linguists: out of the heterogeneous mass of facts, retain only those facts that are related to *one point of view*. This supplementary condition often makes interpretation easier and clearer. In practice, this requirement leads us to identify several groups of variables, some of which have an *active* role in the construction of typologies, while others have the role of *illustrative* variables (also known as supplementary variables).

The difference between analyzed variables and illustrative variables is a fundamental one. We have already encountered this in previous chapters. The ultimate location of a variable that did not participate in the analysis is, in a sense, a validity check; since it did not contribute to a principal axis, interpretation of its correlation with the axis is all the more significant.

Of course, the illustrative variables of one analysis may become the active variables of another analysis, provided they are a homogeneous group of variables; then the formerly active variables become supplementary. This process sometimes yields a more complete interpretation

Thus we may have the following information on a set of individuals:

1. A group of demographic characteristics.
2. A group of variables about attitudes toward work.

By projecting the questions of the second group on the two-dimensional space resulting from the analysis of the first group, we obtain for each attitude variable an analysis of the demographic characteristics of the individuals. It is not difficult to see that the two approaches may lead to complementary results, because the analyses of groups 1 and 2 are likely to lead to two very different typologies of the same individuals. This point was discussed in the Chapter IV example (Section IV.6.2.d).

VII.2. WHAT CAN WE EXPECT FROM MULTIVARIATE DESCRIPTIVE STATISTICAL ANALYSIS?

Our experience has been mainly in economics, social science, and marketing. Therefore we limit ourselves to discussing applications of the methods in these particular fields. There certainly must exist good opportunities for these methods in the natural sciences, psychology, and other disciplines. It is too early to assess the real impact of MDSA on these fields; our evaluation is thus only a partial and temporary one. However, we make a clear distinction between the technical advantages and the more fundamental advantages of MDSA.

VII.2.1. Technical Advantages

(a) Gain in Productivity in Survey Data Processing. By means of the techniques of MDSA, tasks can be ordered rationally in time, thus avoiding confusion among steps. Most of the steps can be illustrated by maps. Finally, information that was formerly inaccessible now becomes available. MDSA also allows us to perform tests of consistency and error detection, as discussed in the next section. In fact, these procedures simultaneously provide a gain in productivity and an improvement in the quality of the results (see Section IV.6).

(b) Tests of Consistency of Data and Error Detection. Detecting outlying or erroneous values is an ancillary result with which statisticians who use principal axes analysis or related techniques are familiar: outliers are often found on the hyperplanes of the first axes.

On the other hand, we can perform a real *evaluation of the data* by positioning "marker variables" on the maps. Just as we can illustrate a map by adding certain characteristics of the individuals that are intrinsic to the survey, we can also use variables that characterize the *way information is gathered*: individuals interviewed by same interviewer; time of interview; interviewer's comments (properly coded), and so on. We can also perform an evaluation of the questionnaire itself, by looking at the positions of the "nonresponses" with respect to the position of the actual responses.

(c) Construction of Artificial Indices. The first axis is the linear combination of the variables having the most variance, and thus often constitute an excellent index for discriminating among individuals. In many applications, this artificial index has a great deal of descriptive power and a meaningful interpretation: it is, for example, the general aptitude factor found in psychological studies. It also allows us to replace large batteries of nominal variables by metric variables, which are much easier to analyze.

VII.2.2. Fundamental Advantages

(a) New Fields of Observation. The possibility of simultaneously treating numerous pieces of information eliminates the need for trimming down a priori the variables to be analyzed in a data set.

Statisticians have long concerned themselves with numerous observations on a small number of variables, whether the purpose be to validate a particular dependency model, or to test a hypothesis. The statistician's concern has always been with handling too many observations, not too many variables. However, the convergence of certain estimates in factor analysis, when the number of variables increases indefinitely, was already mentioned by Hotelling (1933) (more recently, see Wachter, 1978).

Now that the computational obstacle has been removed, "sampling" can be done on both dimensions. In fact, correspondence analysis, which was initially intended for contingency tables, treats these two dimensions symmetrically. The possibility of exploring the "variable" dimension is an innovation the consequences of which are as yet relatively unexplored.

In a discussion concerning economics, Benzecri (1974) doubts that purely analytical data reduction (i.e., explanation of complex phenomena by simple phenomena) can be possible (as they are in physics), because in economics, as in other branches of social sciences, "The order of the composite phenomenon is worth more than the elementary properties of its components."

Data analysis makes it possible to observe complex multidimensional universes, albeit in a still rudimentary fashion, and to globally treat information that previously had to be partitioned to become analyzable.

(b) New Analytical Tools. Presenting the results as maps is in itself a methodological innovation—although the rules for reading these maps are more complicated than they would appear to be. In fact, common language, by its linear and sequential character, makes it easy to express nonsymmetric relationships such as *implications*, whereas the relationship of covariance, which is symmetric, is more difficult to translate into language that implies a causal relationship.

This is why the two-dimensional pictures that represent the factorial planes are very useful tools for analysis and communication.

It is possible to obtain a general overview of large data matrices that is not purely subjective. Thus two researchers who have collected similar data can, in a few words (most often by examining the first two principal axes), grasp the similarities and differences between the two data matrices.

VII.3.   HOW DO WE EVALUATE THE QUALITY OF THE CONFIGURATIONS?

The results of MDSA raise several questions:

1. Are we really observing something that exists? Do the data have a structure? Or, on the contrary, do random errors or sampling fluctuations alone account for the values obtained for the eigenvalues and the explained variances?

2. Do the variances explained represent a *part of information*?

3. Are the configurations obtained *stable*, given what we know about the precision of the data, the nature of the coding, and the relative importance of the different variables?

The following sections attempt to answer each one of these questions.

VII.3.1.   Hypothesis of Independence

The hypothesis of independence of the rows and columns of a matrix is generally too strict to be realistic. It is highly improbable that a matrix being analyzed would be similar to a matrix of random numbers.

Although it is an extreme case with limited applicability, the hypothesis of independence allows us to define *thresholds of significance* for the eigenvalues and the percentages of explained variance, which can serve as guidelines for the user.

The eigenvalues follow nonparametric distributions in the cases of analysis of ranks (Section I.3.4) and of the correspondence analysis of contingency tables (Chapter II). In these favorable circumstances it was possible to obtain *approximate tabulations*, and to draw summary nomograms. The analysis of ranks is discussed in Lebart (1969).

(a)   The Case of Principal Components Analysis.   In some applications the classical correlation coefficients are replaced by Spearman's rank-order correlation coefficients in principal components analysis (cf. Section I.3.4).

We can calculate the correlation matrix by substituting the initial data matrix **R** with matrix **X**, where x_{ij} is the rank of the ith observation on the jth variable, when the observations are ranked in order of magnitude.

Table 1. Principal components analysis of a matrix of ranks; tabulation of percentages of variance under the hypothesis of independence[a]

Table of Means and Standard Deviations of Percentages of Variance Relative to First Eigenvalue

p		n = 20	40	60	80	100	120
5	μ	33.67	29.60	27.59	26.49	25.70	25.11
	σ	3.70	2.92	2.20	1.82	1.73	1.40
10	μ	23.45	18.62	16.76	16.03	15.22	14.79
	σ	2.50	1.25	1.27	0.98	0.80	0.73
15	μ	19.43	14.97	13.11	12.13	11.51	10.98
	σ	1.87	1.02	0.87	0.71	0.66	0.55
20	μ		12.76	10.96	10.14	9.57	9.06
	σ		0.85	0.69	0.59	0.58	0.45
25	μ		11.24	9.62	8.78	8.26	7.81
	σ		0.79	0.63	0.43	0.41	0.37
30	μ		10.21	8.82	7.97	7.36	6.83
	σ		0.67	0.45	0.38	0.33	0.25
35	μ		9.62	8.16	7.23	6.69	6.28
	σ		0.60	0.50	0.40	0.31	0.25
40	μ			7.48	6.68	6.20	5.76
	σ			0.41	0.32	0.24	0.24
45	μ			7.08	6.28	5.79	5.41
	σ			0.30	0.27	0.21	0.23

Table of Means and Standard Deviations of Percentages of Variance Relative to Second Eigenvalue

p		n = 20	40	60	80	100	120
5	μ	25.36	23.70	22.93	22.52	22.21	22.03
	σ	2.78	1.51	1.46	1.10	0.92	0.85
10	μ	18.19	15.62	14.64	13.95	13.64	13.34
	σ	1.46	1.09	0.83	0.70	0.49	0.51
15	μ	15.79	12.67	11.48	10.83	10.43	9.99
	σ	1.29	0.86	0.60	0.51	0.47	0.43
20	μ		11.06	9.66	9.10	8.64	8.18
	σ		0.65	0.49	0.40	0.32	0.33
25	μ		9.97	8.64	7.92	7.55	7.14
	σ		0.48	0.37	0.32	0.31	0.25
30	μ		9.04	7.92	7.16	6.65	6.31
	σ		0.38	0.35	0.31	0.28	0.17
35	μ		8.61	7.27	6.56	6.13	5.82
	σ		0.46	0.33	0.24	0.20	0.24
40	μ			6.77	6.15	5.71	5.35
	σ			0.34	0.21	0.22	0.19
45	μ			6.44	5.80	5.34	5.04
	σ			0.28	0.25	0.19	0.17

[a]p = number of variables; n = number of observations; μ = mean; σ = standard deviation.

Under the hypothesis of independence, the distribution of the Spearman's coefficient depends only on the number of observations n. Similarly, the distribution of the eigenvalues of the correlation matrix of ranks depends only on the parameters n and p, where p is the number of variables.

It is possible to construct an approximate tabulation by *simulation*. We thus obtain an idea of the "degree of significance" of the different percentages of variance under the hypothesis where the p variables are independent by pairs, and have continuous distributions (regardless of the shape of these distributions).

This does not entirely solve the problem of the number of significant axes to retain in the analysis, because the eigenvalues are not independent. However these results are useful as guidelines.

Table 1 shows means and standard deviations of the percentages of variance explained by the first two eigenvalues, as a function of the number of variables and the number of observations. The values corresponding to intermediate dimensions are evaluated by interpolation. In these tables, each (n, p) pair corresponds to 70 simulated analyses established with pseudo-random permutations of ranks.

Let us give, first, an example of complete results (of which the tables are extracts) for the dimensions $n = 60$ and $p = 10$. In Table 2, we note the proximity of the means and medians; the distributions are symmetric, and the usual interval $\{\pm 2\sigma\}$ will give a fairly good approximation of the confidence interval at the 0.95 level.

Figures 1 and 2 show how the percentages change when only one of the matrix's dimensions is varied, whether it be the number of variables (Figure 1) or the number of observations (Figure 2).

(b)   Case of Correspondence Analysis.

Previous Work on the Distribution of Eigenvalues in Correspondence Analysis.   The distribution of eigenvalues extracted from a correspondence analysis under the hypothesis of independence has given rise to a number of

Table 2.

Eigenvalue	Mean	Standard Deviation	Minimum	Median	Maximum
λ_1	16.766	1.268	14.304	16.763	20.826
λ_2	14.639	0.832	13.122	14.672	17.005
λ_3	13.007	0.769	11.325	12.851	14.823
λ_4	11.509	0.523	10.196	11.496	12.593
λ_5	10.215	0.594	8.648	10.181	11.548

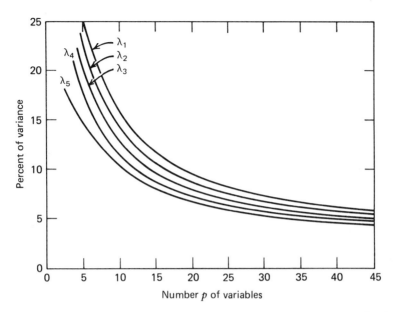

Figure 1. Principal components analysis of a table of ranks: means of percentages of variance relative to the first five eigenvalues. For 100 observations, change as a function of number of variables.

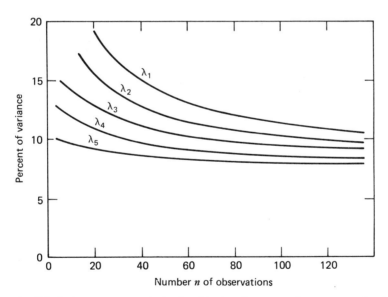

Figure 2. Principal components analysis of a table of ranks: means of percentages of variance relative to the first five eigenvalues. For 15 variables, change as a function of number of observations.

169

erroneous publications. Thus in Kendall and Stuart (1961) the eigenvalues are said to follow the chi-squared distribution, as is the total variance.

Lancaster (1963) refuted this result by showing that the mathematical expected value of the first eigenvalue is always greater than the values derived from Kendall and Stuart's assertions.

References concerning other approximations can be found in the work of Kshirsagar (1972), where it is suggested that eigenvalues, being canonical correlation coefficients calculated on binary variables (cf. Section III.2.5) might follow a distribution very close to that of these same coefficients calculated on Gaussian variables. Simulations have shown that this approximation is unsatisfactory.

We see that the distribution of the eigenvalues can be approximated by that of the eigenvalues of a matrix whose distribution is known (Wishart's matrix). We verify the quality of the approximation by comparing the results of simulations to certain existing tables.

The probability density of the eigenvalues extracted from a Wishart matrix was formulated by Fisher (1939), Girshick (1939), Hsu (1939), and Roy (1939), and then by Mood (1951). The proof is found in Anderson (1958).

Integration of this rather complex density function has given rise to several publications; among them are Pillai (1965), and Krishnaiah and Chang (1971), based on the work of the physicist Mehta (1967).

Tables of thresholds corresponding to the two extreme eigenvalues were published: by Choudary Hanumara and Thompson (1968) for matrices whose smaller dimension p is less than 10; by Pillai and Chang (1970) and by Clemm, Krishnaiah, and Waikar (1973) for $p \leq 20$. As a matter of fact, it can be shown (Lebart, 1975b, 1976; Corsten, 1976; O'Neil, 1978) that the

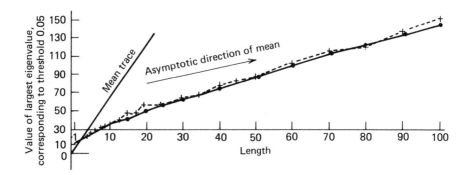

Figure 3. Quality of approximation for first eigenvalue $p = 7$, $n = 1, 2, \ldots, 100$. Dashed line —simulated values (100 simulations/point; solid line—theoretical values (base = 1000).

distribution is related to the Wishart distribution in the following sense: If λ_α is the αth eigenvalue produced by the correspondence analysis of table **K** of order (n, p), with total sum k, then the distribution of $k\lambda_\alpha$ is approximately that of the αth eigenvalue of a Wishart matrix with parameters $n - 1$ and $p - 1$ ("Fisher-Hsu" law).

Figure 3 shows the agreement between the theoretical distribution resulting from the above approximation and the empirical distribution obtained by simulation (100 analyses per point), in the case of matrices for which $p = 7$ and n varies from 1 to 100. To show the result, we use an approximation analogous to that which is made in the establishment of the chi-squared distribution relating to a contingency table (see Section VII.4, Appendix 1).

Independence of the Percentages of Variance and of the Trace (Total Variance). Let us call t the sum of the nontrivial eigenvalues,

$$t = \sum_{\alpha=1}^{p-1} \lambda_\alpha$$

We also define the percentage of variance t_α,

$$t_\alpha = \frac{\lambda_\alpha}{t}$$

If k denotes the sum of all the cells of the (n, p) table **K**, it is well known that kt is nothing but the classical chi-square with $(p - 1)(n - 1)$ degrees of freedom.

The following property holds for the Fisher–Hsu law: the percentages of variance $t_1, t_2, \ldots, t_{p-1}$ are independent of the trace t (see Section VII.4, Appendix 1). This property is valid in the case of correspondence analysis, for which the Wishart distribution is only an approximate distribution (extensive simulations undertaken for constructing the charts have allowed us to verify this independence).

Thus even if the trace does not allow the independence hypothesis to be rejected (the usual chi-squared test), the first percentage of variance can nevertheless be significantly high: correspondence analysis can be used even for tables for which the chi-squared value does not indicate that they are very rich in information.

Conversely, nonsignificant percentages of variance may correspond to a significantly large trace. Although the independence hypothesis is rejected, correspondence analysis is perhaps not the best tool in such a case to describe the dependence between the rows and columns of the table.

Construction of Approximate Nomograms. Figure 4 shows the median values of the largest eigenvalue for the sizes $p = 6, 8, 10, 20, 30, 40, 50$, and

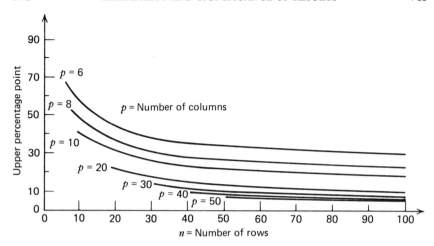

Figure 4. Upper percentage point (0.05) of the largest eigenvalue.

$n = 10$ to 100. The estimates of the values of the percentages of the explained variance corresponding to the first eigenvalue appear on Figure 5 (for a threshold of 0.05). The extremities of the curves (points $(6, 6)$, $(8, 8)$, $(10, 10)$) were established with the help of 1000 simulations (instead of 100 for other points) to clarify their shape.

More detailed information about the construction of these nomograms (particularly on the means used to generate pseudorandom matrices) is

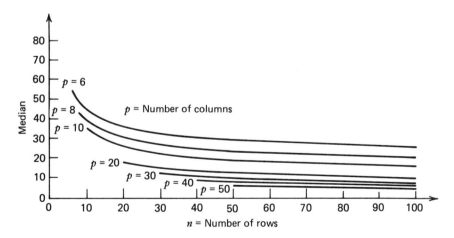

Figure 5. Median values of percentage of variance of largest eigenvalue.

found in Lebart (1975b). The simulation program generates a pseudorandom contingency table with given theoretical marginals and a given total base, using the normal approximation of the multinomial. (Considerable experience has shown that it gives, at a very low cost, comparable results to a procedure using the "exact" multinomial scheme, as far as the eigenvalues are concerned.) We have published approximate tables for contingency tables of maximum size (50, 100), relating to the first five eigenvalues and corresponding rates of variance (estimates of the means, standard deviations, and unilateral 0.05 level for these quantities). Figure 5 summarizes a part of the results for the first eigenvalue. We read, for instance, in Figure 5, that, for a (10, 10) table, the first eigenvalue can reach values higher than 45% of the variance in 5% of the cases (in the hypothesis of independence of the rows and columns of the table).

VII.3.2. Percentage of Variance and Information

Besides the correspondence analysis of *contingency* tables, the use of percentages of variance is very limited. A few counterexamples show us that these coefficients are not suitable for characterizing the quality of a description.

(a) Case of Binary Coding. We have seen (Section IV.2.2) that analyzing two variables with binary coding can give rise to percentages of explained variance that are much smaller than those derived from the equivalent analysis of a contingency table in which the variables are cross-tabulated. Thus percentages of explained variance give an extremely conservative idea of the proportion of "information" that is being represented.

A transformation of the eigenvalues and of the trace (total variance) is needed to make them more meaningful (Benzecri, 1979).

(b) Case of Analysis of Matrix Associated with a Symmetric Graph. In several instances, an accurate analytical calculation is possible without a computer. Then it is interesting to analyze the variations in the mapping as a function of the different coding of the matrix.

Let us examine, for example, the case of the analysis of a *simple cycle*. We designate by n the number of vertices of the graph. The transition equation is written (see Section II.2)

$$\tfrac{1}{2}\mathbf{M}\varphi = \varepsilon(\varphi)\sqrt{\lambda}\,\varphi \tag{1}$$

where $\mathbf{M}$ is the matrix associated with the graph, and $\varepsilon(\varphi) = 1$ or -1 according to the parity of the factor (Benzecri, 1973).

Matrix $\mathbf{M}$ has only two nonzero elements (equal to 1) in each row and column. For $n = 5$, we have, for example,

$$\mathbf{M} = \begin{bmatrix} 0 & 1 & 0 & 0 & 1 \\ 1 & 0 & 1 & 0 & 0 \\ 0 & 1 & 0 & 1 & 0 \\ 0 & 0 & 1 & 0 & 1 \\ 1 & 0 & 0 & 1 & 0 \end{bmatrix} \tag{2}$$

The preceding equation is written, for $1 < j < n$,

$$\tfrac{1}{2}(\varphi(j-1) + \varphi(j+1)) = \varepsilon(\varphi)\sqrt{\lambda}\,\varphi(j) \tag{3}$$

The solutions of this classical type of equation with finite differences are, accounting for the limiting conditions,

$$\varphi_\alpha(j) = \cos\left(\frac{2j\alpha\pi}{n}\right) \quad \text{and} \quad \psi_\alpha(j) = \sin\left(\frac{2j\alpha\pi}{n}\right) \tag{4}$$

These are the jth components of the two factors associated with the double eigenvalue:

$$\lambda_\alpha = \cos^2\left(\frac{2\alpha\pi}{n}\right) \tag{5}$$

In the plane of the first two axes we obtain the parametric equation of a circle, and therefore a satisfactory reconstitution of the structure of which matrix $\mathbf{M}$ represents a particular coding.

The trace of the matrix to be diagonalized is written

$$\mathrm{tr}\left(\tfrac{1}{4}\mathbf{M}^2\right) = \frac{n}{2} \tag{6}$$

The percentage of explained variance corresponding to axis α is thus

$$\tau_\alpha = \frac{2}{n}\cos^2\left(\frac{2\alpha\pi}{n}\right) \tag{7}$$

The apparently paradoxical result is as follows: the variance explained by the subspace that "reconstitutes" the initial structure can be made as small as needed, provided a long enough cycle is chosen (if $n = 10^3$, $\tau_1 \simeq 2 \times 10^{-3}$).

(c) Effects of the Choice of Variables. If a table of n rows and p columns is incremented with q additional columns consisting of random numbers, the normed principal components analysis of the new table with

$p + q$ columns yields the same first axes (if they predominate) as the analysis of the initial table. However, the percentages of explained variance are smaller (because the trace that was equal to p is now equal to $p + q$). Yet the part of information accounted for by the axes naturally stays the same.

In practice, the situation is analogous when the potential number of variables is very large (for example in studies of animal or vegetable species in ecological work). A certain amount of care in the choice of data to collect, called for by homogeneity and exhaustivity requirements, should help to avoid these problems. But the statistician does not always have control over data collection, or sufficient knowledge of the field of application; on the other hand, the criteria for choosing variables are too qualitative and too general to rigorously define an optimal table among all possible tables.

Like coding procedures, the actual choice of variables therefore has more of an influence on the percentages of explained variance than on the principal axes.

(d) Conclusion: Which Information? The preceding counterexamples show that the percentages of explained variance are extremely conservative measures of the quality of an analysis. This is in contrast to multiple correlation coefficients, for example, which are generous measures of the quality of a regression. The initial raw information is not an adequate frame of reference; thus we are often not justified in referring to percentages of variance explained as "parts of information."

It can be shown (see Section VII.5, Appendix 2) that the information theory of Shannon-Wiener (Kullback, 1959), does not allow us to use the percentages of explained variance as a measure of the degree of "nonsphericity" of a configuration of points.

Jeffrey's (1946) *divergence* allows us to express the distance between the hypothesis of independence and the general case as a function of the eigenvalues. Unfortunately, it involves small eigenvalues, whereas correspondence analysis (or related scaling methods) retains only the large ones.

The divergence between the two hypotheses is particularly great in the case where some eigenvalues are close to zero. In the framework of information theory, an infinitely small eigenvalue has a far greater determining role than, for example, three eigenvalues that explain 80% of the total variance.

In fact, as a filter in a process of communication, data analysis has the effect of increasing the *practical value* of the information at the expense of a loss of raw information that may be considerable. But this notion of practical value (Brillouin, 1959) is foreign to the classical theory of information.

As is suggested by Thom (1974), it would be wise to replace the word *information* with the words *form* or *pattern* during an observation process. The best validation criterion is to verify the *stability of the patterns* obtained from the analysis, that is, the stability of the subspace spanned by the first principal axes.

VII.3.3. Stability of the Patterns

Calculations of stability and sensitivity are probably the most convincing validation procedures. These calculations basically consist of verifying the stability of the configurations obtained after modifying the initial matrix in various ways.

What are the various elements that can influence the quality of the results of a principal axes analysis? We can name four:

1. Measurement errors.
2. Choice and weight of variables.
3. Coding of variables.
4. Choice of individuals (or observations), sampling fluctuations.

Each one of these sources of disturbance produces alterations in the initial data matrix, which should not affect the configurations if they are stable. In some cases, a single simulation may be enough, since the purpose is to verify the stability of the initial configuration.

(a) Measurement Errors. The order of magnitude of these errors and their approximate distribution in the population must be specified by the user as a function of his or her own knowledge of the field under study. For example, in the classical case of ordinal responses of the type: "completely disagree," "somewhat disagree," "agree somewhat," "completely agree," we can assume that there is one chance out of two that the respondent answered exactly the way he or she felt, and one chance out of four (except at the extremes) that the answer category was right next to the way the respondent felt.

Computational programs generally allow us to simulate a great variety of situations whose analytical interpretation would be impossible. Because of this, the hypotheses we test may be much better adapted to real situations and to users' actual problems than are classical hypotheses of mathematical statistics. On the other hand, a certain amount of programming is required in order to perform these validations.

(b) Choice and Weight of Variables. This problem arises when the statistician is able to "sample" within the variable space, which is not always the case. Criteria of homogeneity and exhaustivity can only provide

a general framework. "Random samplings" may be performed among the variables, to test the sensitivity of the results with respect to the composition of the variable set.

The problem of variable weights most often occurs in principal components analysis (or in correspondence analysis of binary data, not of counts).

In order to demonstrate stability with respect to the weighting system of the variables, the following transformation may be performed: the initial analysis is done on the standardized variables (standard deviations of 1); then the standard deviations are, for example, expanded to between 1 and 2 (using a pseudorandom generator), and a new, nonnormed analysis is performed on the resulting covariance matrix.

(c) Coding of the Variables. By coding, we mean the preliminary, controlled transformation of the raw data before performing a multidimensional analysis. We feel that this is a basically *empirical* operation, because it is inextricably linked to the contents of the data. Like data analysis itself, coding is intended to increase the *practical value* of the data. The purpose is not to make the data more easily understood by the analyst, but to make it better adapted to the method (this is true for principal axis analysis as well as for classification).

Let us take an example that shows these various aspects of coding. We project, onto a configuration of socioadministrative variables, a new supplementary variable: "time spent during outside work by the woman."

The histogram of this variable is strongly bimodal (Figure 6) because almost half of women do not work out of the home.

The coding problem here is the choice of partitions: choosing the number of classes, and their boundaries. It is clear that the numerical variable, time spent at work, is not usable as such in principal components analysis. In fact, since the calculations only include second-order moments, they are best suited to variables whose marginal distributions are unimodal and relatively symmetric, if not perfectly normal.

Additionally, these calculations favor linear relationships among variables, a peculiarity that is hard to accept in the present case.

It would seem appropriate, a priori, to isolate housewives (doing zero hours of outside work) into one class, and women with exceptionally long work hours into another class. There is no "blind" algorithm that can create a partitioning that satisfies these two requirements with one histogram (an algorithm such as Fisher's, 1958, which gives "exact optimal" partitions into k classes, cannot, with less than five classes, isolate housewives into one class).

Working empirically, we constructed five classes from a detailed histogram, and obtained, in the plane of the first two principal axes, the result

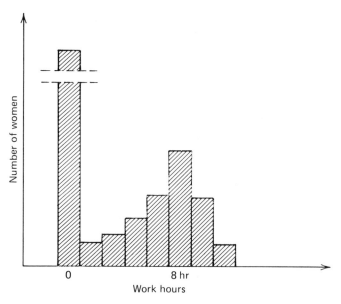

Figure 6.   Histogram of female work hours.

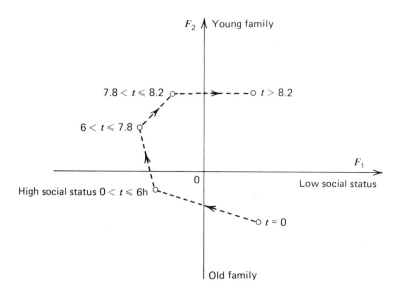

Figure 7.   Trajectory of female work hours (axes 1 and 2).

shown schematically in Figure 7 (where the interpretation of the axes is briefly sketched).

The relationship with the first axis is nonlinear, and would not have been seen if the variable had been positioned as a numerical variable, without dividing it into classes, nor if only two classes had been used. Thus, thanks to the loss of information by dividing into classes, and to constructing classes based on purely semantic class limits, this descriptive method gives the most meaningful information on the data being analyzed.

Thus coding is not simply a formality before analysis: it requires a simultaneous knowledge of the data (from elementary descriptive statistics) and of the method.

On the other hand, coding can be a source of disturbance in the case of ratings, scales, and rankings (for example, in analysis of ranks or preferences). It is important, then, to verify that the configurations obtained are resistant to monotone transformations with a high degree of distortion (logs, exponentials, etc.), to be sure that the order of the ratings is more important than the particular metric properties of the scale used.

Finally, in every case, it is interesting to try "minimal" coding, that is the simplest coding that conserves the observed configurations. Let us give an example: a principal components analysis was performed on a table of individual consumer expenses, creating a certain typology of expenses. This analysis was done again, this time by coding the expenses into "1" if they were incurred, regardless of their size. The resulting typology of expenses was very similar to the preceding one. Thus the interpretation of the first analysis can be modified by this new result, which emphasizes the importance of the existence of certain types of expenses, independently of the size of the expense. (cf. Jousselin, 1972)

(d) Sampling Fluctuations and Bootstrap. Two types of stability calculations can be undertaken as in Section VII.3.3b above. First of all, the *weighting* of the individuals can be modified. For example, the disturbances caused by giving each individual a random weight drawn uniformly between 0 and 2 can be studied (then the expected value of the total weight stays equal to the number of observations).

Another procedure consists of splitting the sample to test the stability of the results. General statistical methods of the jackknife (cf. Miller, 1974) and bootstrap (Efron, 1979) types gives nonparametric results that are solely conditioned by the data itself. The statistical importance of these methods, and the ease with which they are executed, probably justify the increased amount of computation they imply.

The jackknife procedure is not particularly appropriate in the case of sampling fluctuations of the eigenvalues, whose distribution depends in a complex way on both parameters p and n. This drawback does not exist in the case of the bootstrap method, which does not alter the values of these parameters. Bootstrap can be used to evaluate the degree of separation of two or more consecutive eigenvalues. Nonparametric confidence intervals can be drawn; the axes characterized by nonoverlapping intervals are worth interpreting. But as shown in Section VII.3.2, the eigenvalues are not always a reliable measure of the quality of the results.

A more interesting use of the bootstrap involves studying the reliability of the coordinates of points on the principal axes; this may lead one to compare the points' respective positions on the graphic display (see Section VII.3.4).

The required amount of computation could be very large, since these techniques require numerous diagonalizations of a matrix that may be large.

A specific procedure can be used to save computer time, using some of the results in Chapter VI. To avoid an entire diagonalization at each step of the bootstrap, a direct reading can be performed, starting from the observed eigenvalues (which is much faster than starting with a random configuration).

In any case, the amount of computation remains considerable.

(e) Asymptotic Results.   Certain specific simulations can give information concerning sampling fluctuations. In the case of principal components analysis (on a matrix of correlations or covariances), the sampling may be done with the normal distribution, using the covariance matrix computed on the observations. The eigenvalues and eigenvectors (thus the coordinates) calculated from the simulated tables can then be compared to the results of the analysis.

When only the eigenvalues are of interest, the simulations can be avoided by using the asymptotic results of Anderson (1963). If the theoretical eigenvalues l_α are distinct, then the eigenvalues of the empirical covariance matrix $\mathbf{C}$ follow asymptotically the normal distribution with the mean l_α and the variance $2l_\alpha^2/(n-1)$, where n is the size of the sample. We can deduce the approximate 0.95 confidence intervals:

$$l_\alpha \in \left[\lambda_\alpha \left(1 - 1.96 \sqrt{\frac{2}{n-1}} \right); \ \lambda_\alpha \left(1 + 1.96 \sqrt{\frac{2}{n-1}} \right) \right] \tag{8}$$

The size of the interval gives an idea of the stability of the eigenvalue with respect to fluctuations due to sampling (supposedly normal). The overlapping of the intervals of two consecutive eigenvalues suggests the equality (or the proximity) of these eigenvalues. The corresponding axes are then defined within a rotation. Thus the user can restrict the effort of interpretation to the subspace corresponding to the first eigenvalues with good separation.

Generalizations of the asymptotic results to the nonnormal case can be found (Davis, 1977), but their use is not practical.

Even though Anderson's result concerns eigenvalues of covariance matrices, its use can often be extended to correlation matrices. The simulations show that the intervals obtained are generally conservative: the percentage of coverage of the real value is most often greater than the confidence threshold used. In every case, the asymptotic nature of the results and the underlying hypothesis of normal distribution of the observations imply that the results should be considered as indicative, not imperative.

Figure 8 shows Anderson's asymptotic confidence intervals at the level 0.95 for the first five eigenvalues of the correlation matrix of the example of Chapter I: only the first two eigenvalues appear well separated. To verify the procedures we performed 1000 simulations of correlation matrices. The eigenvalues of the analysis are therefore theoretical values, which the 1000 Anderson intervals should cover in 95% of the cases. The number of covering intervals is given for each eigenvalue in Table 3.

As far as the eigenvalues of the covariance matrix of the same data matrix are concerned, the eigenvalues are better separated, and the coverage rates of the confidence intervals are more regular.

Note 1. In the case of correspondence analysis, the elements of the asymptotic distribution of the eigenvalues are found in O'Neill (1978, 1981).

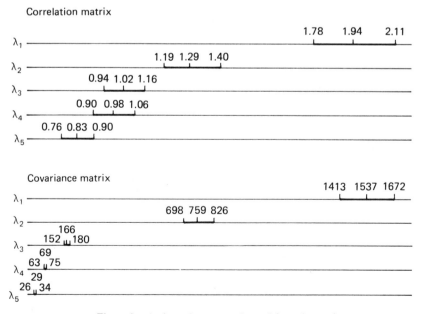

Figure 8. Anderson's asymptotic confidence intervals.

Table 3.   Number of covering intervals (threshold 0.95)

Number of Simulations	λ_1	λ_2	λ_3	λ_4	λ_5	λ_6	λ_7	λ_8
Correlation matrix								
100	99	98	98	99	97	97	94	93
500	496	485	490	496	488	490	459	465
1000	990	975	979	986	975	978	921	934
Covariance matrix								
100	97	94	93	94	92	90	97	90
500	479	469	471	474	475	469	473	461
1000	954	944	946	953	950	949	955	930

A recent reference concerning the asymptotic study of eigenvectors is found in Tyler (1981), who generalizes the results obtained by Anderson. A general presentation of asymptotic problems is in Muirhead (1982).

Note 2.   The typologies obtained by principal components analysis do not require as strict a representation of the sample as do percentage or mean estimations (obviously, a sample where certain aspects of the parent population are absent cannot give projectable results, even if the configurations obtained are stable). This relative stability with respect to the representation of the sample is a fact stemming from experience. However, a very partial explanation can be given by recalling that the subspaces corresponding to the largest eigenvalues are the most stable with respect to possible disturbances of the matrix to be diagonalized (cf. Wilkinson, 1965; Escofier and Leroux, 1972). Additionally, this matrix itself (for example, the sample correlation matrix in principal components analysis) is less sensitive to sampling fluctuations than the first-order moments (means or percentages).

VII.3.4.   Confidence Areas for Points on Graphical Displays

When a point is the center of gravity of a group of individuals, it can easily be assigned a confidence circle in the subspace spanned by the first axes. The interpretation is as follows: the smaller the radius of the circle with respect to the distance to the center, the more the point is different from the general mean point, and therefore, the more important is the location of this point on the plane (as far as interpreting the axes is concerned).

(a)   The Case of (n, p) Contingency Tables.   Each of the n rows of the contingency table can be represented as a point z_i with p coordinates $z_{ij}(j \leq p)$. Namely, $z_{ij} = f_{ij}/f_i.$. The row z_i is weighted by $f_i.$. The mean-point g of the n profiles $z_1, \ldots, z_n$ is a vector with the coordinates $f._j$ ($j \leq p$)

(see Section II.5.1). The chi-square distance between z_i and g is noted $d(z_i, g)$ with

$$d^2(z_i, g) = \sum_{j=1}^{p} \frac{1}{f_{\cdot j}}(z_{ij} - g_j)^2 \qquad (9)$$

that is,

$$d^2(z_i, g) = \sum_{j=1}^{p} \frac{1}{f_{\cdot j}}\left(\frac{f_{ij}}{f_{i\cdot}} - f_{\cdot j}\right)^2 \qquad (10)$$

If k denotes the total sum of the elements of the contingency table, the quantity $c_i^2 = kf_{i\cdot}.d^2(z_i, g)$ can be written

$$c_i^2 = k \sum_{j=1}^{p} \frac{(f_{ij} - f_{i\cdot}.f_{\cdot j})^2}{f_{i\cdot}.f_{\cdot j}} \qquad (11)$$

The quantity c_i^2 approximately follows a chi-squared distribution with $(p - 1)$ degrees of freedom (d.f.) if the row i is supposed to be filled according to a multinomial distribution, the theoretical probabilities for each cell being defined by the margins. In other words, z_i only differs from g on account of sampling fluctuation.

After the projection onto any two-dimensional subspace containing g (e.g., the subspace of the first two principal axes of correspondence analysis), the squared distance will follow a chi-squared distribution with two d.f. (the idempotent matrix of projection being of rank two).

This leads to a simple procedure to test the significance of the position of certain points on the graphic displays.

We can draw a confidence circle centered at the origin with radius: $r = \sqrt{5.99/kf_{i\cdot}}$. (5.99 is the value given by the tables for $p = 0.05$ and d.f. = 2).

The projection of z_i will fall outside this circle with the probability 0.05, if the ith row of the table is statistically equivalent to the margin.

In practice, instead of drawing concentric circles around the origin, it is clearer and easier to draw them around each point concerned, and look at the position of the origin (see example below).

The reduction of the d.f. from $(p - 1)$ to 2 only holds if the two-dimensional subspace is fixed in advance. However, it remains valid if the points involved z_i do not participate much in the construction of the principal axes (cases of small weighted points, of points close to the origin, or of supplementary points, i.e., points plotted afterwards).

Numerical Application. Let us consider again the contingency table of Chapter II. In addition, the matrix has four illustrative columns.

The total sum k is equal to 3160 (number of respondents). The trace is equal to $t = 0.5922$. Thus $kt = 1713$.

This value is highly significant, for a chi-squared variable with 336 d.f. The hypothesis of independence is rejected, and correspondence analysis allows us to understand why it is rejected, by pointing out the network of relationships between the rows and the columns of the table. For the first axis, the percentage of variance is 36%. The statistical table summarized in Figure 5 gives the value of 20%, corresponding to the confidence level 0.05. Therefore, this axis is significant. (Without knowledge of the conditional distribution of the percentage corresponding to the second axis, it is not possible to easily appreciate its significance.)

Confidence circles have been drawn around the four illustrative columns (Figure 9) that appear to occupy significant positions on the plane. However the point "autonomy," which is as far from the center as the illustrative point "medium size," does not differ significantly from the midpoint (notice that "autonomy" is an active point, but with small absolute contribution, which allows the construction of the confidence area).

(b) Case of a Complete Binary Matrix (Multiple Correspondence Analysis). A complete binary matrix $\mathbf{Z}$, having n rows, is analyzed. A supplementary category j is a column containing n_j values of 1 and $n - n_j$ values of 0. The coordinate of the corresponding point on an axis where the variance is λ is written $\varphi_j = \overline{X}_j / \sqrt{\lambda}$ where $\overline{X}_j$ is the mean of the coordinates of the n_j individuals who chose this category.

Let us suppose as a working hypothesis that the n_j individuals are drawn randomly, without replacement, from the n individuals. Then point j only differs from the origin because of chance fluctuations. Under these conditions the expected value of the coordinate of j on an axis is zero, its variance is $(n - n_j)/n_j(n - 1)$, and the covariance of the coordinates of the point on two distinct axes is zero.

By application of the central limit theorem to the mean variable, the distribution of the coordinate on an axis is approximated by the normal distribution and the coordinates of a point on distinct axes are independent. The square of the distance to the origin in a q-dimensional subspace thus follows approximately a chi-squared distribution with q degrees of freedom.

Thus, for example, j has a position that is different from the origin in the plane spanned by the two first axes if the circle of radius

$$\sqrt{\frac{n - n_j}{n - 1} \frac{5.99}{n_j}} \qquad (12)$$

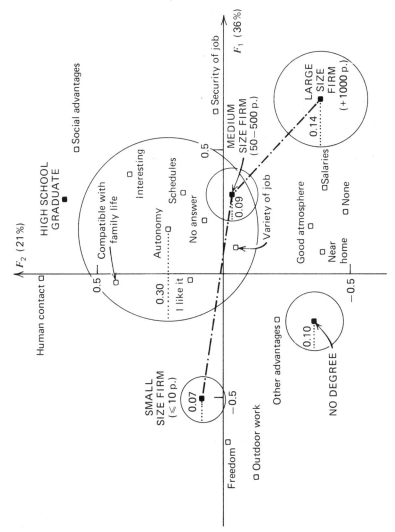

Figure 9. Confidence circles: □—active point; ■—illustrative point.

185

does not contain the origin. The smaller the radius of this circle is with respect to the distance to the origin, the more important is the point's position on the plane. The cumulated squares on the q first axes can be calculated to compare them to the chi-squared thresholds with $1, 2, \ldots, q$ degrees of freedom.

(c)   Case of Principal Components Analysis.   Any category of a nominal variable that is projected as a supplementary element in the space of the first axes of a principal components analysis is defined as the mean point of the individuals it represents. The procedure shown in Section VII.3.4b thus applies equally well in this case.

In the principal components analysis of a correlation matrix, the coordinate $\hat{\varphi}_{j\alpha}$ of variable j on axis α is equal to the correlation coefficient between the n values of variable $j(\mathbf{x}_j)$ and the n coordinates of the individuals on axis $\alpha(\hat{\psi}_\alpha)$:

$$\hat{\psi}_{j\alpha} = \mathrm{cor}(\mathbf{x}_j, \hat{\psi}_\alpha) \tag{13}$$

The significance of the position of an illustrative continuous variable j on axis α is based on this coefficient. It is interpreted with reference to the null hypothesis of independence. Under this hypothesis the correlation has an expected value of zero and a variance of $1/(n-1)$. Thus in a principal plane, the important supplementary variables are those outside a circle that is centered at the origin and whose radius is $2/\sqrt{n-1}$ (approximate level = 0.05).

Note that these correlation coefficients can also be used in multiple correspondence analysis, in order to position the continuous variables as illustrative elements in the principal planes. The coordinate of such a variable on an axis is, as before, its correlation coefficient with the axis. The same confidence circle allows us to select the continuous variables that are useful for interpreting the axes or the planes.

Concerning the analysis of matrices where one of the dimensions is individual cases, this discussion may be summarized as follows:

If the analysis is done on nominal variables (multiple correspondence), or on continuous variables (principal components), there are two types of supplementary elements: the illustrative *continuous variables*, which are important outside the central circle, and the illustrative *category-points*, which are computed as centers of gravity, around which specific confidence circles are drawn.

VII.4.   APPENDIX 1: EIGENVALUES AND PERCENTAGES OF VARIANCE IN CORRESPONDENCE ANALYSIS (INDEPENDENCE HYPOTHESIS)

VII.4.1.   An Approximation of the Distribution of the Eigenvalues

We use here the same notation as in Chapter II. k_{ij} is the general term of the contingency table $\mathbf{K}$ with n rows and p columns. We note again that

$$k = \sum_i \sum_j k_{ij} \quad \text{and} \quad f_{ij} = \frac{k_{ij}}{k} \tag{14}$$

If p_{ij} designates the probability corresponding to cell (i, j), and if the theoretical margins are noted $p_i.$ and $p._j$, the hypothesis of independence of the rows and columns is written

$$p_{ij} = p_i.p._j \tag{15}$$

Thus k_{ij} is one of the np components of a multinomial vector, with

$$E(k_{ij}) = kp_i.p._j \tag{16}$$

(E is the expected value).

An approximation is made that is analogous to that which is made in establishing the chi-squared distribution for a contingency table: k is supposed to be large enough to allow the utilization of the normal approximation of the multinomial distribution. On the other hand, we note that the observed marginals, $f_i.$ and $f._j$, can be substituted for the theoretical marginals $p_i.$ and $p._j$ *without neglecting the constraints that are implied by this substitution*.

Let us designate by $\mathbf{h}$ the vector with np components such that

$$h_{ij} = \frac{\sqrt{k}\,(f_{ij} - f_i.f._j)}{\sqrt{f_i.f._j}} \tag{17}$$

Under the preceding conditions, this vector of $\mathbb{R}^{np}$ has a normal distribution with $E(h_{ij}) = 0$ for all i and j.

The general term of its covariance matrix is

$$V_h(i, j; i', j') = \delta_{ij,i'j'} - (f_i.f._jf_{i'}.f._{j'})^{1/2} \tag{18}$$

with $\delta_{ij,i'j'} = 1$ if $i = i'$ and $j = j'$; otherwise 0.

Let us construct a (p, p) orthogonal matrix $\mathbf{A}$ such that its first column's jth element is $f_{\cdot j}^{1/2}$ (for all $j \leq p$), and the other $p - 1$ columns form an orthonormal basis of $\mathbb{R}^p$ with the first column. Similarly, let us construct an (n, n) orthogonal matrix $\mathbf{B}$ such that its first row has $f_{i\cdot}^{1/2}$ for its ith element (for all $i \leq n$), and the other $n - 1$ rows form an orthonormal basis of $\mathbb{R}^n$ with the first row.

The (np, np) matrix $\mathbf{B} \otimes \mathbf{A}'$, the direct product, or Kronecker product of the matrices $\mathbf{B}$ and $\mathbf{A}'$, is also orthogonal.

We have the equations

$$\sum_j \sqrt{f_{\cdot j}}\, h_{ij} = 0 \qquad \text{for all } i \tag{19}$$

$$\sum_i \sqrt{f_{i\cdot}}\, h_{ij} = 0 \qquad \text{for all } j \tag{20}$$

$$\sum_m b_{rm}\sqrt{f_{m\cdot}} = 0 \qquad \text{for } 1 < r \leq n \tag{21}$$

$$\sum_k a_{ks}\sqrt{f_{\cdot k}} = 0 \qquad \text{for } 1 < s \leq p \tag{22}$$

From these equations, we deduce that vector $\mathbf{y}$ of $\mathbb{R}^{np}$ such that

$$\mathbf{y} = \mathbf{B} \otimes \mathbf{A}'\mathbf{h} \tag{23}$$

has only $(n - 1)(p - 1)$ nonzero components. We have

$$y_{rs} = 0 \qquad \text{if } r = 1 \text{ or if } s = 1$$

The covariance matrix of $\mathbf{y}$ is

$$\mathbf{V}_y = (\mathbf{B} \otimes \mathbf{A}')\mathbf{V}_h(\mathbf{B}' \otimes \mathbf{A}) \tag{24}$$

For every pair of nonzero components we have

$$\mathbf{V}_y(r, s; r', s') = \delta_{rr'}\delta_{ss'} \tag{25}$$

Let $\mathbf{Y}$ be the (n, p) matrix defined by

$$\mathbf{Y} = \mathbf{BHA}$$

where $\mathbf{H}$ is the (n, p) matrix whose general term is h_{ij}. The first row and the first column of $\mathbf{Y}$ are zeroes.

The elements of the $(n - 1, p - 1)$ submatrix $\hat{\mathbf{Y}}$, formed with the non-zero elements of $\mathbf{Y}$ are thus distributed independently according to the standardized normal distribution. The matrix

$$\mathbf{S} = \hat{\mathbf{Y}}'\hat{\mathbf{Y}}$$

is thus distributed according to Wishart's distribution with the parameters $(n - 1)$ and $(p - 1)$ (cf. Anderson, 1958).

$\mathbf{S}$ has the same nonzero eigenvalues as $\mathbf{Y}'\mathbf{Y}$, that is, as $\mathbf{A}'\mathbf{H}'\mathbf{H}\mathbf{A}$; that is, again the same eigenvalues as $\mathbf{H}'\mathbf{H}$, since $\mathbf{A}$ is orthogonal.

Note that this implies that $\operatorname{tr}\mathbf{H}'\mathbf{H}$ is a chi-square with $(n - 1)(p - 1)$ degrees of freedom. But

$$\operatorname{tr}\mathbf{H}'\mathbf{H} = \frac{k\sum_i \sum_j (f_{ij} - f_i f_{\cdot j})^2}{f_i f_{\cdot j}} \tag{26}$$

We are dealing with the usual chi-squared test on contingency tables.

Note that the symmetric matrix $\mathbf{S}^*$ that is diagonalized during the correspondence analysis of matrix $\mathbf{K}$ is the matrix

$$\mathbf{S}^* = \frac{1}{k}\mathbf{H}'\mathbf{H}$$

VII.4.2. Independence of the Percentages of Variance and of the Trace (Total Variance) (cf. Bartlett (1951), see also Davis (1972)).

The density of the joint distribution of the eigenvalues $\lambda_1, \lambda_2, \ldots, \lambda_p$ of a Wishart matrix has the classical form

$$w(\Lambda) = C(n, p)\Pi\lambda_\alpha^{(n-p-1)/2}\exp\left\{-\frac{1}{2}\sum\lambda_\alpha\right\}\Pi_{\alpha<\beta}(\lambda_\alpha - \lambda_\beta) \tag{27}$$

with

$$C(n, p) = \left(\frac{\pi^{p/2}}{2^{np/2}}\right)\Pi\Gamma\left(\frac{n + 1 - \alpha}{2}\right)\Gamma\left(\frac{p + 1 - \alpha}{2}\right) \tag{28}$$

If we suppose

$$\lambda_\alpha = z\tau_\alpha \qquad \text{for } \alpha < p \tag{29}$$

and

$$\lambda_p = (1 - \tau_1 - \tau_2 - \cdots - \tau_{p-1})z \tag{30}$$

then z is the trace of the matrix. It is easy to find a factorization of the density (the jacobian of this transformation is z^{p-1}):

$$w(\Lambda) = w_1(z)w_2(\tau_1,\ldots,\tau_{p-1}) \qquad (31)$$

with

$$z = \Sigma\lambda_\alpha \qquad (32)$$

and

$$w_1(z) = \frac{1}{2\Gamma\left(\dfrac{np}{2}\right)}\left(\frac{z}{2}\right)^{(np/2)-1}\exp\{-z/2\} \qquad (33)$$

which is the chi-squared distribution with np degrees of freedom.

The percentages of variance $\tau_1, \tau_2,\ldots,\tau_{p-1}$ are therefore independent of the trace z.

VII.5   APPENDIX 2: INFORMATION AND EIGENVALUES

For these calculations we use the concept of *divergence* of Jeffreys (1946), which allows us to measure the distance between two hypotheses H_1 and H_2 in the case of a multidimensional sample x, stemming from one of the two schemes relative to normal distributions in $\mathbb{R}^p$:

$$H_1 = \text{hypothesis of independence}\begin{cases} \text{theoretical mean } = \mu_1 \\ \text{theoretical covariance matrix } = \sigma^2\mathbf{I} \end{cases}$$

$$H_2 = \text{general case}\begin{cases} \text{theoretical mean } = \mu_2 \\ \text{theoretical covariance matrix } = \mathbf{S} \\ \text{(supposedly nonsingular)} \end{cases}$$

The divergence allows us to express the distance between the hypotheses H_1 and H_2 as a function of the eigenvalues of $\mathbf{S}$, and we see that it involves small eigenvalues whereas principal components analysis only retains the large ones.

For two hypotheses H_1 and H_2 that can give rise to a sample x, the divergence $J(H_1, H_2)$ is defined as the difference

$$J(H_1, H_2) = \int\log\frac{P(H_1|\mathbf{x})}{P(H_2|\mathbf{x})}d\nu_1(\mathbf{x}) - \int\log\frac{P(H_2|\mathbf{x})}{P(H_1|\mathbf{x})}d\nu_2(\mathbf{x}) \qquad (34)$$

v_1 and v_2 are the measures associated with the hypotheses H_1 and H_2, and $P(H_i|\mathbf{x})$ is the conditional probability that H_i is true, given that $\mathbf{x}$ is known. In the case of continuous densities $f_1(\mathbf{x})$ and $f_2(\mathbf{x})$, we have

$$J(H_1, H_2) = \int (f_1(\mathbf{x}) - f_2(\mathbf{x}))\log\frac{f_1(\mathbf{x})}{f_2(\mathbf{x})}\,d\mathbf{x} \tag{35}$$

The probability density of sample $\mathbf{x}$ is written, for a theoretical covariance matrix $\mathbf{S}_i$, and for a mean vector $\boldsymbol{\mu}_i$:

$$f_i(\mathbf{x}) = \frac{1}{|2\pi\mathbf{S}_i|^{1/2}}\exp - \left\{\tfrac{1}{2}(\mathbf{x} - \boldsymbol{\mu}_i)'\mathbf{S}_i^{-1}(\mathbf{x} - \boldsymbol{\mu}_i)\right\} \tag{36}$$

Hence

$$\log\frac{f_1(\mathbf{x})}{f_2(\mathbf{x})} = \frac{1}{2}\log\frac{|\mathbf{S}_2|}{|\mathbf{S}_1|} - \frac{1}{2}\mathrm{tr}(\mathbf{S}_1^{-1}(\mathbf{x} - \boldsymbol{\mu}_1)(\mathbf{x} - \boldsymbol{\mu}_1)')$$

$$+ \tfrac{1}{2}\mathrm{tr}(\mathbf{S}_2^{-1}(\mathbf{x} - \boldsymbol{\mu}_2)(\mathbf{x} - \boldsymbol{\mu}_2)') \tag{37}$$

The first term of $J(H_1, H_2)$ is $I(1; 2)$, the mean information given by sample $\mathbf{x}$ under the hypothesis H_1, for discriminating in favor of H_1 against H_2 (cf. Kullback, 1959).

$$I(1; 2) = \int f_1(\mathbf{x})\log\frac{f_1(\mathbf{x})}{f_2(\mathbf{x})}\,d\mathbf{x} \tag{38}$$

$$= \frac{1}{2}\log\frac{|\mathbf{S}_1|}{|\mathbf{S}_2|} + \frac{1}{2}\mathrm{tr}(\mathbf{S}_1(\mathbf{S}_2^{-1} - \mathbf{S}_1^{-1}))$$

$$+ \tfrac{1}{2}\mathrm{tr}(\mathbf{S}_2^{-1}(\boldsymbol{\mu}_1 - \boldsymbol{\mu}_2)(\boldsymbol{\mu}_1 - \boldsymbol{\mu}_2)') \tag{39}$$

We have written: $\mathbf{x} - \boldsymbol{\mu}_2 = \mathbf{x} - \boldsymbol{\mu}_1 + \boldsymbol{\mu}_1 - \boldsymbol{\mu}_2)$. We have

$$J(H_1, H_2) = I(1; 2) + I(2; 1) \tag{40}$$

$$= \tfrac{1}{2}\mathrm{tr}((\mathbf{S}_1 - \mathbf{S}_2)(\mathbf{S}_2^{-1} - \mathbf{S}_1^{-1}))$$

$$+ \tfrac{1}{2}\mathrm{tr}((\mathbf{S}_1^{-1} + \mathbf{S}_2^{-1})(\boldsymbol{\mu}_1 - \boldsymbol{\mu}_2)(\boldsymbol{\mu}_1 - \boldsymbol{\mu}_2)') \tag{41}$$

We are interested in the case where

$$\boldsymbol{\mu}_1 = \boldsymbol{\mu}_2 \qquad \mathbf{S}_1 = \mathbf{I} \qquad \text{and} \qquad \mathbf{S}_2 = \mathbf{S} \tag{42}$$

We note in abbreviated form that

$$J(\mathbf{I}, \mathbf{S}) = \tfrac{1}{2}\mathrm{tr}\big((\mathbf{I} - \mathbf{S})(\mathbf{S}^{-1} - \mathbf{I})\big) = \tfrac{1}{2}\mathrm{tr}(\mathbf{S} + \mathbf{S}^{-1}) - p \qquad (43)$$

or, by bringing in the eigenvalues of $\mathbf{S}$,

$$J(\mathbf{I}, \mathbf{S}) = \frac{1}{2}\left(\sum_{\alpha=1}^{p} \lambda_\alpha + \sum_{\alpha=1}^{p} \frac{1}{\lambda_\alpha} \right) - p \qquad (44)$$

If the total theoretical variances are equal under hypotheses H_1 and H_2, we have

$$\sum_{\alpha=1}^{p} \lambda_\alpha = p \qquad (45)$$

The only variable term in $J(\mathbf{I}, \mathbf{S})$ is therefore

$$\sum_{\alpha=1}^{p} \frac{1}{\lambda_\alpha} \qquad (46)$$

We see that the divergence between the two hypotheses is particularly great in the case where some of the eigenvalues of $\mathbf{S}$ are close to zero.

CHAPTER VIII

A Computer Program: Correspondence Analysis for Large Matrices

The goal of this chapter is to provide readers with the material necessary to perform a correspondence analysis on their own data. A user's guide, a FORTRAN listing, test data, and an example of output corresponding to the test data are included.

VIII.1.   MAIN FEATURES OF THE PROGRAM

(a)   The program is written in portable FORTRAN IV; it can be run on most medium-size or large computers equipped with 32 bit words, or with equivalent capabilities.

(b)   It comprises two steps derived from a more general program named SPAD, which contains many more procedures for analyzing large survey files. As such, it may be viewed as a relatively simplified version, with an abbreviated dictionary for identifying variables and individuals. It can be simplified even further (see Section VIII.3, technical remarks).

(c)   It can handle large data matrices, with no practical limitations on the numbers of rows (the data matrix is never stored in the central memory). If required, it can perform an out-of-core diagonalization, using some of the results of Chapter VI. In this case, the number of columns can also be very large. For example, a (5000, 500) matrix can easily be handled under this option (see user's guide to determine the amount of memory required).

(d)   The program is self-contained, and provided with the necessary numerical and graphical procedures.

(e)   This program can, of course, be used to process *binary tables* as well as *contingency tables*. Its performance in the case of large binary complete tables cannot compete with a multiple correspondence analysis program, working directly on the reduced coding matrix (see Chapter IV).

VIII.2.   PARAMETERS

CORAN performs the computations and outputs the results of a correspondence analysis; in this version, which is designed to handle large data matrices, eigenvalues may be computed in main memory or by direct reading. In order to use CORAN, the following parameters are assembled:

CARD 1:   TITLE
Study title in 80 characters.

CARD 2:   5 PARAMETERS, in format 5 I4.
1.   IEXA   Number of rows in data file.
2.   NQEXA   Number of columns in data file. (The row identifier is not counted as one of the columns.)
3.   NVIDI   Length of row identifier; this must be a multiple of 4 characters. Maximum, NVIDI = 15, corresponding to 60 characters (the first 4, or 8, or 12 characters appear on the graphs). Such an identifier is required, and must be at the beginning of each row.
4.   LFMT   Number of format cards (if 0, the default value is LFMT = 1).
5.   MODIG   Mode of defining row selection. There are two formats for selecting and defining rows for analysis. If MODIG = 0: all the rows are active.
If MODIG = 1: there are one or more selection cards. (Some rows are either supplementary or ignored.)

CARD(S) 3:   COLUMN IDENTIFIERS
4 character identifiers are read in fixed format (20A4) for the NQEXA original columns (that is, before columns are selected and shifted around).

CARD(S) 4:   COLUMN SELECTION
Codes are read in format 80 I1 for the NQEXA columns of the data matrix, such that:
0 = column eliminated from analysis
1 = active column
2 = illustrative column

CARD(S) 5: ROW SELECTION
When MODIG = 0: there are no row selection cards. All rows are active.
When MODIG = 1: codes are read in 80 I1 for all IEXA rows, such that:
0 = row eliminated from analysis
1 = active row
2 = illustrative row

CARD(S) 6: FORMAT FOR READING DATA
The format is written in parentheses on LFMT cards. The format begins with NVIDI × A4. The remainder is read in real (F) format (even if integer values).

CARD(S) 7: THE DATA
The data is read according to the previous format. There are IEXA rows, each row is NVIDI + NQEXA long.

CARD 8: 7 PARAMETERS, in format 7 I4:
1. NFAC   Number of principal coordinates to compute
2. LIST3   Parameter for printing row information:
           0 = no printing
           1 = coordinates and contributions of rows are printed (column information is always printed)
3. NGRAF   Number of graphs (NGRAF ≤ 10). In this version, the principal planes are successively: (1, 2); (2, 3); (3, 4); etc. All points—rows, columns, active and supplementary—are shown.
4. NPAGE   Number of pages for the width of each graph (NPAGE ≤ 8).
5. NLIGN   Number of lines per graph. Recommended:
           NLIGN = 60 × NPAGE − 2
           If NLIGN = 0; both axes are scaled identically.
6. JBASE   Dimension of the space for approximation in the case of direct reading:
           If JBASE = 0: diagonalization takes place in main memory (usual choice).
           If JBASE = 1: the default value is JBASE = NFAC + 3.
           Otherwise: JBASE = dimension of space for approximation.
7. NITER   (Only if JBASE ≠ 0): number of iterations in direct reading. If NITER = 0: the default is NITER = 8.

VIII.2.1. Storage Requirements for CORAN

The main program contains a vector **Q**, of length MOTS, which controls the allocation of memory for the entire program. The length of the vector **Q** is a function of the dimensions of the data matrix to be analyzed, and the specific parameters selected (in particular, the mode of diagonalization). The length of **Q** can be calculated using the formula below.

We note   NVAR is the total number of active and illustrative columns.

ITOT is the total number of rows in the data file.

NACT is the number of active columns.

NFAC is the number of principal axes requested.

JBASE is the parameter defined above.

$$N = \max(\text{NVAR} \times \text{NFAC}; \; 3 \times (\text{ITOT} + \text{NVAR}))$$

FIRST CASE.   JBASE = 0.

$$\text{MOTS} \geq \text{NACT}^2 + 2 \times \text{ITOT} + 9 \times \text{NVAR} + 2 \times N$$

SECOND CASE.   JBASE ≠ 0.

$$M = \max(N; \; \text{NACT} \times \text{JBASE})$$

$$\text{MOTS} \geq \text{JBASE}^2 + 2 \times \text{ITOT} + 9 \times \text{NVAR} + 3 \times M$$

Example.   If the data matrix is (1000, 200), and if we wish to extract six factors, all rows and columns being active:

$$\text{NACT} = \text{NVAR} = 200$$

$$\text{NFAC} = 6$$

$$\text{ITOT} = 1000$$

Thus

$$N = 3500$$

In the first case

$$\text{MOTS} \geq 51,000$$

In the second case (with the default value JBASE = NFAC + 3)

$$\text{MOTS} \geq 7481$$

VIII.3. TECHNICAL REMARKS

VIII.3.1. Principal Computational Steps

(a) Case of Direct Reading (JBASE ≠ 0). The subroutine LECDI performs a calculation through the decomposed iterated power algorithm (see Chapter VI). Each iteration involves reading row by row the data matrix (subroutine PUISC). A shift of origin (see Section VI.2) is performed. Subroutine CPROJ computes the matrix to be diagonalized after projection onto the subspace extracted by PUISC. A classical diagonalization is then performed on this small matrix (VPROP).

(b) Classical Case (JBASE = 0). The symmetrized matrix (see Section II.5.2) is computed in subroutine MEMOI; this matrix is centered in order to eliminate the trivial eigenvalue 1, then diagonalized (VPROP).

VIII.3.2. Possible Simplification of the Program

(Numbers refer to the listing of Section VIII-7.) The user wishing to create a shorter version of this program can skip parts of it.

(a) If the user has a large machine, or moderate-size data matrices, the direct reading procedure can be ignored. In this case, the user must suppress the calling statement of subroutine LECDI (statements 149 to 151) and the subroutines LECDI, GSMOD, PUISC, SEN3A, and CPROJ.

(b) The user can also skip the command for graphical displays. (Note that all the eigenvalues, coordinates on principal axes, and identifiers can be saved on the tape NGUS = 11.) Statements 215 to 234 must be suppressed and subroutines HPLAN, EPUR4, and BORNS must be ignored.

(c) The user can also replace the classical diagonalization subroutine that is provided (VPROP, TRIDI) by the one available in his or her own library.

VIII.4. COMMENTS ON THE OUTPUT EXAMPLE

This example is performed on the same data set as the example of Chapter II. It differs from it in the following respects: To show a more general configuration of parameters, different sets of active and supplementary elements are chosen. As a consequence, the numerical results are similar, but not identical to those of Chapter II. In particular, the two principal axes are oriented in opposite directions (in any case, these directions are arbitrary).

The two steps SELEC and CORAN (extracted from the more general package SPAD) communicate through the file NLEG (= 13 in the example); that accounts for the "check-list" of reread parameters the reader can find on the output after the parameters of CORAN.

```
ANALYSIS OCCUPATIONS / JOB ADVANTAGES
   26   22    6    1    1
VARIFREEHUMASCHESALASECUCOMPINTENEARATMOSOCIAUTOLIKEOTHENONEOUTDNOANNODEGRADSMAL
MEDILARG
1111111111111011022222
11111111111112111111111012
(6A4,F1.0,22F4.0 )
```

Occupation		VARI	FREE	HUMA	SCHE	SALA	SECU	COMP	INTE	NEAR	ATMO	SOCI	AUTO	LIKE	OTHE	NONE	OUTD	NOAN	NODE	GRAD	SMAL	MEDI	LARG
FARM*FARMING/FISHING	4	189	0	3	2	2	9	3	12	2	1	4	11	15	12	8	1	93	18	187	2	0	
FAR2*FARM/FOOD INDUSTRY	1	13	3	10	17	12	4	1	8	3	5	1	9	5	11	0	0	32	15	12	40	11	
ENER*ENERGY/MINES	1	9	1	0	4	13	0	2	2	6	2	0	4	3	6	0	0	9	8	6	17	6	
STEE*STEEL	5	5	2	9	18	5	3	2	6	1	5	1	2	3	22	1	0	29	8	4	30	27	
CHEM*CHEMICAL/GLASS/OIL	2	7	1	4	15	5	1	1	6	1	2	0	1	0	0	1	0	13	13	3	14	25	
WOOD*WOOD/PAPER	2	5	0	4	1	0	0	2	2	0	1	1	6	0	3	2	0	10	5	6	11	14	
AUT *AUTO/AVIATION/SHIP	2	3	1	8	16	17	1	7	7	1	4	3	1	24	0	1	0	25	4	7	19	61	
TEXT*TEXTILE/LEATHER	3	18	8	6	16	5	0	8	13	4	2	3	6	2	26	2	0	28	4	24	38	6	
PHAR*PHARMACY INDUSTRY	3	7	3	6	16	6	0	2	6	3	3	0	2	1	26	0	0	11	6	16	16	9	
MANU*MANUFACTURING	0	18	1	12	31	7	0	6	19	11	3	4	10	8	35	6	0	51	26	32	40	30	
CONS*CONSTRUCTION	2	63	2	9	31	9	4	7	8	10	0	1	14	1	7	3	2	69	28	74	49	8	
FOOD*FOOD/GROCERY	2	43	16	7	6	4	7	1	9	0	3	6	0	4	26	3	0	26	9	81	13	1	
SBUS*SMALL BUSINESS	8	95	23	15	15	2	13	7	9	8	3	5	13	8	18	3	3	34	62	163	21	3	
MBUS*MISCEL.BUSINESS	5	32	9	9	17	5	4	7	9	4	2	3	8	1	18	3	3	33	29	67	29	37	
ADMI*ADMINISTRATIVE SER.	8	26	10	24	24	80	10	11	7	11	8	2	6	4	16	2	3	35	78	53	79	10	
TELE*TELECOMMUNICATION	1	7	2	11	11	3	1	6	2	3	1	1	2	3	5	0	4	5	10	20	10	1	
SO.S*SOCIAL SERVICES	4	10	10	8	2	1	6	4	19	2	3	2	1	2	1	0	1	6	25	11	10	29	
HE.S*HEALTH SERVICES	3	31	16	15	11	19	5	10	10	10	4	3	7	5	5	5	1	17	78	32	48	14	
TEAC*TEACHING/RESEARCH	3	33	27	31	9	18	27	24	3	4	43	24	18	2	11	1	5	12	188	62	83	14	
TRAN*TRANSPORTATION	2	19	2	12	12	21	2	3	4	5	5	8	3	3	13	1	3	23	10	20	35	10	
BANK*INSURANCE/BANKING	8	12	4	8	13	21	1	10	2	5	7	1	6	1	10	0	1	7	44	25	24	2	
DOME*DOMESTIC WORKERS	0	8	0	4	5	2	2	4	5	2	6	1	3	0	1	1	3	23	1	34	3	11	
O.SE*OTHER SERVICES	8	35	14	13	16	10	6	25	10	6	4	9	11	4	14	1	1	33	82	79	35	11	
PRIN*PRINTING/PUBLISHING	2	13	2	14	5	8	0	10	0	8	3	2	5	4	11	2	0	17	19	27	24	1	
PRIV*PRIVATE SERVICES	0	26	9	3	12	5	8	4	4	3	10	3	8	3	8	3	2	21	16	77	9	5	
NO A*NO ANSWER	0	14	15	3	4	4	3	1	1	1	5	1	3	1	3	3	0	7	15	19	11		

```
   6    1    2    1   58    0    0
```

VIII.6. OUTPUT EXAMPLE

```
STEP    ** SELEC **

TITLE=ANALYSIS OCCUPATIONS / JOB ADVANTAGES

  IEXA= 26    NQEXA= 22    NVIDI= 6    LFMT = 1    MODIG= 1

NAMES OF COLUMNS

VARI FREE HUMA SCHE SALA SECU COMP INTE NEAR ATMO SOCI AUTO LIKE OTHE NONE OUTD NOAN NODE GRAD SMAL
MEDI LARG

SUMMARY OF SELECTION

  TYPE 1   NUMBER OF VARIABLES   15

  TYPE 2   NUMBER OF VARIABLES    5

INDICATOR VECTOR OF   22   ELEMENTS IN GROUPS OF 10
1111111111 1110110222 22

SUMMARY OF SELECTION

  TYPE 1   NUMBER OF VARIABLES   23

  TYPE 2   NUMBER OF VARIABLES    2

INDICATOR VECTOR OF   26   ELEMENTS IN GROUPS OF 10
1111111111 1112211111 111012

FORMAT

(6A4,F1.0,22F4.0 )


                              END OF READING AND SELECTION
```

PARAMETER CARD FOR   CORAN

           NFAC =   6   LIST3= 1   NGRAF=   2   NPAGE= 1   NLIGN= 58   JBASE= 0   NITER= 0

  INPUT FILE =   13 (NLEG)   ANALYSIS OCCUPATIONS / JOB ADVANTAGES

OUTPUT FILE   = 14 (NSAV)

OUTPUT FILE   = 11 (NGUS)

RE-READING PARAMETERS ON THE FILES
           ICARD= 23   ITOT = 25   NACT = 15   NVAR = 20   NVIDI= 6

MEMORY USAGE   YOU HAVE RESERVED 2000   YOU NEED   649

        EIGENVALUES

SUM OF THE EIGENVALUES   0.52593672

HISTOGRAM OF THE FIRST EIGENVALUES

       EIGENVALUE   PERCENTAGE   PERCENTAGE
                                    CUM.

    1   0.21108344   40.13   40.13   ***
    2   0.12640679   24.03   64.17   ***
    3   0.06108136   11.61   75.78   *********************
    4   0.03939711   7.49   83.27   **************
    5   0.02358575   4.48   87.76   ********
    6   0.01786559   3.40   91.16   ******
    7   0.01308984   2.49   93.64   *****
    8   0.00997749   1.90   95.54   ****
    9   0.00763348   1.45   96.99   ***
   10   0.00636142   1.21   98.20   ***
   11   0.00393719   0.75   98.95   **
   12   0.00341867   0.65   99.60   **
   13   0.00167373   0.32   99.92   *
   14   0.00042438   0.08   100.00   *

200

COORDINATES AND CONTRIBUTIONS OF THE COLUMNS

NAMES	MASSES	DIST.	F1	F2	F3	F4	F5	F6	F1	F2	F3	F4	F5	F6	F1	F2	F3	F4	F5	F6
				COORDINATES						ABSOLUTE CONTRIBUTIONS						SQUARED CORRELATIONS				
VARI	0.028	0.48	-0.08	0.02	0.07	-0.25	-0.09	-0.58	0.1	0.0	0.2	4.5	0.9	51.9	0.01	0.00	0.01	0.13	0.02	0.69
FREE	0.250	0.53	0.72	0.03	-0.08	0.02	-0.01	-0.01	61.8	0.1	2.4	0.3	0.1	0.1	0.98	0.00	0.00	0.00	0.00	0.00
HUMA	0.053	0.80	-0.02	-0.70	0.28	-0.39	-0.20	0.04	0.0	20.5	6.7	20.1	8.8	0.4	0.00	0.61	0.10	0.19	0.05	0.00
SCHE	0.075	0.26	-0.33	-0.18	0.08	0.03	-0.20	-0.01	3.8	2.0	0.8	0.2	13.1	0.1	0.41	0.13	0.02	0.00	0.16	0.00
SALA	0.101	0.38	-0.36	0.43	0.11	-0.07	0.06	0.03	6.1	14.5	1.9	1.3	1.5	0.6	0.34	0.48	0.03	0.01	0.01	0.00
SECU	0.071	0.86	-0.55	0.00	-0.71	0.02	-0.23	0.05	10.2	0.0	58.1	0.1	15.4	1.0	0.35	0.00	0.58	0.00	0.06	0.00
COMP	0.043	0.59	0.06	-0.51	0.38	0.20	-0.12	0.03	0.1	8.8	9.9	4.6	2.5	0.2	0.01	0.44	0.24	0.07	0.02	0.00
INTE	0.053	0.56	-0.34	-0.43	-0.15	-0.19	0.38	-0.11	3.0	7.9	1.9	5.1	33.0	3.4	0.21	0.33	0.04	0.07	0.26	0.02
NEAR	0.056	0.30	-0.11	0.39	0.15	-0.09	0.02	0.18	0.3	6.7	2.0	1.1	0.1	9.7	0.04	0.50	0.07	0.02	0.00	0.10
ATMO	0.031	0.54	-0.23	0.36	0.39	0.09	-0.07	0.03	0.7	3.2	7.6	0.6	0.6	0.2	0.09	0.24	0.28	0.01	0.01	0.00
SOCI	0.039	1.28	-0.52	-0.68	0.11	0.67	0.13	0.00	5.0	14.5	0.7	45.3	2.8	0.0	0.21	0.36	0.01	0.35	0.01	0.00
AUTO	0.023	0.45	-0.20	-0.25	-0.29	-0.08	0.37	-0.13	0.4	1.1	3.2	0.4	12.8	2.2	0.09	0.14	0.19	0.02	0.30	0.04
LIKE	0.062	0.19	-0.05	-0.12	-0.01	-0.18	0.17	0.27	0.1	0.7	0.0	5.1	7.3	25.2	0.01	0.08	0.00	0.17	0.15	0.38
NONE	0.109	0.36	-0.25	0.48	0.12	0.11	0.01	-0.09	3.3	19.5	2.6	3.1	0.1	4.8	0.18	0.62	0.04	0.03	0.00	0.02
OUTD	0.005	3.37	1.43	0.21	-0.49	0.80	0.22	0.05	4.9	0.2	2.0	8.2	1.0	0.1	0.61	0.01	0.07	0.19	0.01	0.00

SUPPLEMENTARY ELEMENTS

NAMES	MASSES	DIST.	F1	F2	F3	F4	F5	F6	F1	F2	F3	F4	F5	F6	F1	F2	F3	F4	F5	F6
NODE	0.222	0.23	0.17	0.38	0.09	0.10	0.02	0.05	0.0	0.0	0.0	0.0	0.0	0.0	0.12	0.63	0.03	0.04	0.00	0.01
GRAD	0.254	0.65	-0.24	-0.69	0.02	0.04	0.17	-0.02	0.0	0.0	0.0	0.0	0.0	0.0	0.09	0.74	0.00	0.00	0.05	0.00
SMAL	0.372	0.41	0.52	-0.07	0.11	-0.09	-0.01	-0.03	0.0	0.0	0.0	0.0	0.0	0.0	0.67	0.01	0.03	0.02	0.00	0.00
MEDI	0.220	0.27	-0.41	0.00	-0.02	0.08	-0.06	0.07	0.0	0.0	0.0	0.0	0.0	0.0	0.61	0.00	0.00	0.02	0.01	0.02
LARG	0.103	1.68	-0.75	0.43	-0.26	-0.02	0.08	0.11	0.0	0.0	0.0	0.0	0.0	0.0	0.33	0.11	0.04	0.00	0.00	0.01

COORDINATES AND CONTRIBUTIONS OF THE ROWS

NAMES	MASSES	DIST.	F1	F2	F3	F4	F5	F6	F1	F2	F3	F4	F5	F6	F1	F2	F3	F4	F5	F6
				COORDINATES						ABSOLUTE CONTRIBUTIONS						SQUARED CORRELATIONS				
FARM*FARMING	0.095	1.40	1.14	0.10	-0.21	0.18	0.06	-0.00	59.0	0.7	6.6	8.0	1.6	0.0	0.93	0.01	0.03	0.02	0.00	0.00
FAR2*FARM/FO	0.036	0.25	-0.32	0.17	-0.03	0.06	-0.13	0.25	1.7	0.8	0.1	0.4	2.7	12.2	0.41	0.11	0.00	0.02	0.07	0.24
ENER*ENERGY/	0.017	0.88	-0.22	0.13	-0.79	0.10	-0.10	0.12	0.4	0.2	17.1	0.5	0.6	1.4	0.05	0.02	0.71	0.01	0.01	0.02
STEE*STEEL	0.032	0.59	-0.48	0.42	0.29	0.15	-0.09	-0.21	3.5	4.6	4.5	1.9	1.2	8.0	0.39	0.31	0.15	0.04	0.02	0.07
CHEM*CHEMICA	0.020	0.46	-0.33	0.33	0.02	-0.06	0.03	0.08	1.1	1.8	0.0	0.2	0.1	0.8	0.24	0.24	0.00	0.01	0.00	0.02
WOOD*WOOD/PA	0.009	0.65	-0.00	-0.02	0.36	0.19	-0.18	-0.28	0.0	0.0	1.9	0.8	1.1	3.8	0.00	0.00	0.20	0.06	0.05	0.12
AUT *AUTO/AV	0.037	0.59	-0.61	0.32	-0.24	0.04	0.06	-0.03	6.6	2.9	3.5	0.1	0.6	0.1	0.64	0.17	0.10	0.00	0.01	0.00
TEXT*TEXTILE	0.040	0.36	-0.19	0.47	0.16	0.04	0.11	-0.02	0.7	7.1	1.6	0.2	2.2	0.1	0.10	0.62	0.07	0.00	0.04	0.00
PHAR*PHARMAC	0.018	0.42	-0.22	0.21	0.38	-0.01	-0.04	-0.12	0.4	0.6	4.3	0.0	0.1	1.5	0.12	0.11	0.35	0.00	0.00	0.04
MANU*MANUFAC	0.054	0.51	-0.33	0.52	0.19	-0.04	0.14	0.20	2.7	11.4	3.1	0.2	4.6	12.4	0.21	0.52	0.07	0.00	0.04	0.08
CONS*CONSTRU	0.019	0.51	0.12	0.36	0.04	0.03	0.09	-0.06	0.5	7.8	0.2	0.2	2.5	1.5	0.08	0.70	0.01	0.01	0.04	0.02
SBUS*SMALL B	0.083	0.28	0.45	-0.11	0.16	-0.15	-0.11	-0.04	7.8	0.8	3.5	4.9	4.0	0.9	0.71	0.05	0.09	0.08	0.04	0.01
MBUS*MISCEL.	0.045	0.10	0.04	0.13	0.21	-0.10	-0.08	-0.04	0.0	0.6	3.2	1.1	1.1	0.3	0.01	0.16	0.42	0.09	0.06	0.01
TELE*TELECOM	0.021	0.77	-0.45	-0.14	-0.54	-0.04	-0.29	-0.05	2.1	0.3	10.1	0.1	7.6	0.3	0.26	0.03	0.37	0.00	0.11	0.00
SO.S*SOCIAL	0.020	0.80	-0.00	-0.55	0.42	-0.34	-0.34	-0.15	0.0	4.8	5.8	5.9	9.7	2.6	0.00	0.38	0.22	0.14	0.14	0.03
HE.S*HEALTH	0.062	0.36	-0.14	-0.31	-0.21	-0.35	0.13	0.21	0.6	4.8	4.5	19.4	4.3	14.7	0.05	0.27	0.12	0.34	0.04	0.12
TEAC*TEACHIN	0.094	0.81	-0.29	-0.76	0.11	0.36	0.05	0.06	3.7	42.5	2.0	31.2	1.0	1.7	0.10	0.71	0.02	0.16	0.00	0.00
TRAN*TRANSPO	0.036	0.47	-0.31	0.20	-0.37	0.16	-0.35	0.02	1.7	1.2	8.3	2.3	19.1	0.1	0.21	0.09	0.30	0.05	0.26	0.00
BANK*INSURAN	0.039	0.51	-0.43	-0.05	-0.43	-0.09	-0.01	-0.30	3.5	0.1	11.9	0.8	0.0	20.2	0.37	0.00	0.37	0.01	0.00	0.18
DOME*DOMESTI	0.020	0.75	-0.12	0.29	0.40	0.39	-0.06	0.10	0.1	1.3	5.3	7.7	0.3	1.3	0.02	0.11	0.21	0.20	0.01	0.01
O.SE*OTHER S	0.066	0.24	-0.14	-0.28	-0.03	-0.15	0.27	-0.18	0.7	4.2	0.1	3.9	20.6	12.2	0.09	0.33	0.01	0.10	0.31	0.14
PRIV*PRIVATE	0.038	0.15	0.05	-0.14	0.11	-0.20	0.11	0.05	0.0	0.6	0.8	3.7	1.9	0.6	0.02	0.13	0.08	0.26	0.08	0.02

SUPPLEMENTARY ELEMENTS

NAMES	MASSES	DIST.	F1	F2	F3	F4	F5	F6	F1	F2	F3	F4	F5	F6	F1	F2	F3	F4	F5	F6
ADMI*ADMINIS	0.090	1.06	-0.48	-0.06	-0.75	0.03	-0.41	0.01	0.0	0.0	0.0	0.0	0.0	0.0	0.21	0.00	0.53	0.00	0.16	0.00
NO A*NO ANSW	0.022	0.86	0.05	-0.58	0.15	-0.43	-0.19	0.19	0.0	0.0	0.0	0.0	0.0	0.0	0.00	0.38	0.03	0.21	0.04	0.04

AXIS 1 /HORIZONTAL   AXIS 2 /VERTICAL

ATTENTION

THE POINTS BELOW WERE MORE THAN   2.5 STD.DEVIATIONS FROM CENTER
THEY HAVE BEEN BROUGHT BACK TO PERIPHERY OF GRAPH

```
* * * * * * * * * * * * * * * * * * * * * * * * * * * * *
*FARM*FARMING *        1.14356       *       0.09690       *
*OUTD         *        1.43197       *       0.21027       *
* * * * * * * * * * * * * * * * * * * * * * * * * * * * *
```

```
 0.518  I  --------------------------MANU*MANUFAC----------- .  -------------------------------------------------------------  I
 0.496  I                             NONE                   :                                                                I
 0.474  I                                 TEXT*TEXTILE       :                                                                I
 0.451  I                                                    :                                                                I
 0.429  LARG        STEE*STEEL  SALA                         :                                                                I
 0.407  I                                   NEAR             :           NODE                                                 I
 0.384  I                             ATMO                   :                                                                I
 0.362  I                                                    :           CONS*CONSTRU                                         I
 0.339  I                        CHEM*CHEMICA                :                                                                I
 0.317  I       AUT *AUTO/AV                                 :                                                                I
 0.295  I                                     DOME*DOMESTI   :                                                                I
 0.272  I                                                    :                                                                I
 0.250  I                                                    :                                                                I
 0.228  I                              PHAR*PHARMAC          :                                                                I
 0.205  I                           TRAN*TRANSPO             :                                                                I
 0.183  I                           FAR2*FARM/FO             :                                                                I
 0.161  I                                                    :                                                                I
 0.138  I                              ENER*ENERGY/          .  MBUS*MISCEL.                                                  I
 0.116  I                                                    :                                                           OUTD I
 0.094  I                                                    :                                                           FARM I
 0.071  I                                                    :                                                                I
 0.049  I                                                    :                                                                I
 0.027  I                                         VARI       :                                                           FREE I
 0.004  I    .  . SECU     MEDI     .   .   .   .   .   +  .   .   .   .   .   .   .   .                                       I
-0.018  I                                                WOOD*WOOD/PA                                                         I
-0.040  I                   BANK*INSURAN                     :                                                                I
-0.063  I                   ADMI*ADMINIS                     :                                              SMAL              I
-0.085  I                                                    :                                                                I
-0.107  I                                       LIKE         :                               SBUS*SMALL B                    I
-0.130  I                   TELE*TELECOM                      .  PRIV*PRIVATE                                                  I
-0.152  I                                                    :                         FOOD*FOOD/GR                          I
-0.174  I                             SCHE                   :                                                                I
-0.197  I                                                    :                                                                I
-0.219  I                                                    :                                                                I
-0.241  I                               AUTO                 :                                                                I
-0.264  I                               O.SE*OTHER S         :                                                                I
-0.286  I                                                    :                                                                I
-0.308  I                           HE.S*HEALTH              :                                                                I
-0.331  I                                                    :                                                                I
-0.353  I                                                    :                                                                I
-0.375  I                                                    :                                                                I
-0.398  I                                                    :                                                                I
-0.420  I                           INTE                     :                                                                I
-0.442  I                                                    :                                                                I
-0.465  I                                                    :                                                                I
-0.487  I                                                    :  COMP                                                         I
-0.509  I                                                    :                                                                I
-0.532  I                                             SO.S*SOCIAL                                                            I
-0.554  I                                              .  NO A*NO ANSW                                                       I
-0.576  I                                                    :                                                                I
-0.599  I                                                    :                                                                I
-0.621  I                                                    :                                                                I
-0.643  I                                                    :                                                                I
-0.666  I                   SOCI                             :                                                                I
-0.688  I                                     GRAD        HUMA                                                               I
-0.711  I                                                    :                                                                I
-0.733  I                                                    :                                                                I
-0.755  I  --------------------------TEAC*TEACHIN----------- .  -------------------------------------------------------------  I
       -0.745               -0.452               -0.158               0.135                0.429                0.722
```

                    AXIS 2 /HORIZONTAL   AXIS 3 /VERTICAL

                    ATTENTION

THE POINTS BELOW WERE MORE THAN   2.5 STD.DEVIATIONS FROM CENTER
THEY HAVE BEEN BROUGHT BACK TO PERIPHERY OF GRAPH

* *
*ENER*ENERGY/ *   0.13138   *   -0.78926   *
* *

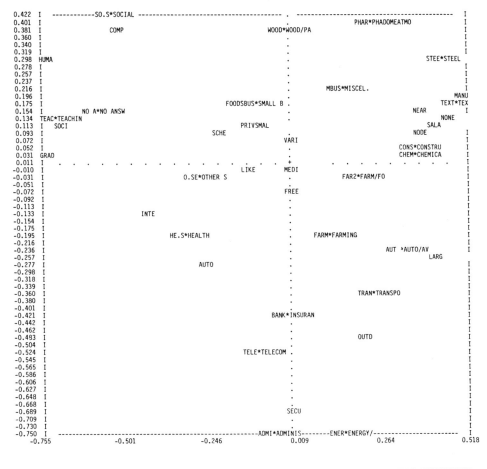

```
 0.422  I  -----------SO.S*SOCIAL ------------------------------------ .  --------------------------------------------  I
 0.401  I                                                              .                                               I
 0.381  I                   COMP                                       .              PHAR*PHADOMEATMO                  I
 0.360  I                                               WOOD*WOOD/PA   .                                               I
 0.340  I                                                              .                                               I
 0.319  I                                                              .                                               I
 0.298  HUMA                                                           .                                STEE*STEEL     I
 0.278  I                                                              .                                               I
 0.257  I                                                              .                                               I
 0.237  I                                                              .                                               I
 0.216  I                                                              .      MBUS*MISCEL.                             I
 0.196  I                                                              .                                        MANU
 0.175  I                                    FOODSBUS*SMALL B .                                          TEXT*TEX
 0.154  I         NO A*NO ANSW                                         .                          NEAR           I
 0.134  TEAC*TEACHIN                                                   .                              NONE
 0.113  I  SOCI                                     PRIVSMAL           .                          SALA
 0.093  I                          SCHE                                .                          NODE            I
 0.072  I                                                 VARI         .                                          I
 0.052  I                                                              .                      CONS*CONSTRU        I
 0.031  GRAD                                                           .                      CHEM*CHEMICA        I
 0.011  I  .    .    .    .    .    .    .    .    LIKE      MEDI      +    .    .    .    .    .    .    .    .    I
-0.010  I                                                              .                                          I
-0.031  I                       0.SE*OTHER S                           .           FAR2*FARM/FO                   I
-0.051  I                                                              .                                          I
-0.072  I                                                 FREE         .                                          I
-0.092  I                                                              .                                          I
-0.113  I                                                              .                                          I
-0.133  I           INTE                                               .                                          I
-0.154  I                                                              .                                          I
-0.175  I                                                              .                                          I
-0.195  I             HE.S*HEALTH                                      .       FARM*FARMING                       I
-0.216  I                                                              .                                          I
-0.236  I                                                              .                      AUT *AUTO/AV        I
-0.257  I                                                              .                            LARG
-0.277  I                  AUTO                                        .                                          I
-0.298  I                                                              .                                          I
-0.318  I                                                              .                                          I
-0.339  I                                                              .                                          I
-0.360  I                                                              .       TRAN*TRANSPO                       I
-0.380  I                                                              .                                          I
-0.401  I                                                              .                                          I
-0.421  I                               BANK*INSURAN                   .                                          I
-0.442  I                                                              .                                          I
-0.462  I                                                              .                                          I
-0.483  I                                                              .           OUTD                           I
-0.504  I                                                              .                                          I
-0.524  I                    TELE*TELECOM .                            .                                          I
-0.545  I                                                              .                                          I
-0.565  I                                                              .                                          I
-0.586  I                                                              .                                          I
-0.606  I                                                              .                                          I
-0.627  I                                                              .                                          I
-0.648  I                                                              .                                          I
-0.668  I                                                              .                                          I
-0.689  I                               SECU                           .                                          I
-0.709  I                                                              .                                          I
-0.730  I                                                              .                                          I
-0.750  I  -----------------------------------------ADMI*ADMINIS--------ENER*ENERGY/-----------------------  I
        -0.755              -0.501              -0.246             0.009              0.264              0.518
```

                        END OF STEP   ** CORAN   **

VIII.7. FORTRAN LISTING

```
C+++++++  CORAN = CORRESPONDENCE ANALYSIS                            1
C                                                                    2
      DIMENSION Q(2000)                                              3
      COMMON /ENSOR/LEC,IMP                                          4
      DATA NLEG/13/,  NGUS/11/ ,NSAV/14/,NBAND/15/                   5
C.....NLEG,NGUS,NSAV,NBAND =4 AUXILIARY FILES.                       6
C.....LEC,IMP =INPUT AND OUTPUT (PRINTER) FILE NUMBERS.             7
      MOTS = 2000                                                    8
      LEC  = 5                                                       9
      IMP  = 6                                                      10
      CALL SELEC ( Q , MOTS ,  NLEG , NBAND )                       11
      CALL CORAN ( Q , MOTS ,  NLEG , NGUS , NBAND , NSAV )         12
      STOP                                                          13
      END                                                           14
                                                                    15
      SUBROUTINE CORAN ( Q, MOTS, NLEG, NGUS, NBAND, NSAV )         16
C * * * * * * * * * * * * * * * * * * * * * * * * * * * * * * * *   17
C  CALLS =      - TITLE *                                     *     18
C               - DCALA - TITLE *                             *     19
C                       - PRECO *                             *     20
C                       - LECDI - SEN3A *                     *     21
C                               - TITLE *                     *     22
C                               - GSMOD *                     *     23
C                               - PUISC - TITLE *             *     24
C                                       - GSMOD *             *     25
C                               - CPROJ - TITLE *             *     26
C                                       - VPROP - TRIDI *     *     27
C                       - MEMOI - VPROP - TRIDI *             *     28
C                       - AFCOB - TITLE *                     *     29
C                               - OPEN *                      *     30
C               - VPOUR *                                     *     31
C               - OPEN *                                      *     32
C               - HPLAN - BORNS *                             *     33
C                       - EPUR4 - BORNS *                     *     34
C INPUT FILE=NLEG   OUTPUT=NGUS  (TEMPORARY=NBAND,NSAV)       *     35
C * * * * * * * * * * * * * * * * * * * * * * * * * * * * * * * *   36
      DIMENSION Q(MOTS), LTITR(20)                                  37
      COMMON  / ENSOR / LEC,IMP                                     38
      DATA LSAV/4HNSAV/, LGUS/4HNGUS/                               39
C.......... READING  OF PARAMETERS FOR   DCALA                      40
      WRITE (IMP,2000)                                              41
      READ(LEC,1000) NFAC,LIST3,NGRAF,NPAGE,NLIGN,JBASE,NITER       42
 1000 FORMAT(2OI4)                                                  43
      WRITE (IMP,2100) NFAC, LIST3, NGRAF, NPAGE, NLIGN, JBASE, NITER  44
      IF (NFAC .LE. 0)             GO TO 50                         45
C.......... READING OF TITLE AND PARAMETERS                         46
      CALL TITLE (1, NLEG)                                          47
      WRITE (IMP,2500) NSAV, LSAV                                   48
      WRITE (IMP,2500) NGUS, LGUS                                   49
      REWIND NLEG                                                   50
      READ (NLEG) (LTITR(L),L=1,20), LLEG,                          51
     1           ICARD, ITOT, NACT, NVAR, NVIDI                     52
      NQFIN = NVAR                                                  53
      NDIM  = NVAR                                                  54
```

```
      NMAX  = NACT                                                    55
      IJTOT = ITOT + NVAR                                             56
      IJDIM = IJTOT + 1                                               57
      IF (NGRAF .EQ. 0)              IJDIM = 1                        58
      IF (JBASE .EQ. 1)             JBASE = NFAC + 3                  59
      IF (JBASE .GT. NACT)          JBASE = 0                         60
      LBASE = JBASE                                                   61
      IF (JBASE .EQ. 0)            LBASE = 1                          62
      IF (NITER .EQ. 0)            NITER = 8                          63
      KB    = JBASE                                                   64
      IF (JBASE .EQ. 0)              KB = NACT                        65
      WRITE (IMP,2200)  ICARD, ITOT, NACT, NVAR, NVIDI               66
C......... MEMORY RESERVATION                                         67
      NS    = 1                                                       68
      NX    = NS                                                      69
      ND    = NS + NMAX*KB                                            70
      ND1   = NS + IJDIM                                              71
      IF( ND1 .GT. ND)             ND = ND1                           72
      NPJ   = ND + NACT                                               73
      NCO   = NPJ + NVAR                                              74
      NCOOR = NCO + NVAR                                              75
      NBZ   = NCOOR                                                   76
      NID   = NCOOR                                                   77
      NU    = NCOOR + NDIM*NFAC                                       78
      NU2   =NCOOR + NACT * JBASE                                     79
      IF( NU2  .GT.  NU )          NU = NU2                           80
      NU3   = NCOOR + IJDIM*NVIDI                                     81
      IF( NU3 .GT. NU )           NU = NU3                            82
      NW    = NU + NQFIN                                              83
      NIDJ  = NW + NFAC                                               84
      NAD   = NIDJ + NQFIN                                            85
      NV    = NAD + JBASE*JBASE                                       86
      NY    = NV                                                      87
      NFIN  = NV + NQFIN                                              88
      NFIN2 = NV + IJDIM                                              89
      IF(NFIN2  .GT. NFIN)           NFIN = NFIN2                     90
      WRITE (IMP,2300) MOTS, NFIN                                     91
      IF ( NFIN .GT. MOTS )          GO TO 50                         92
      CALL DCALA (NFAC,LIST3,JBASE,LBASE,NITER,NGRAF,NLIGN,NPAGE,     93
     1  KB,    NDIM,NMAX, NQFIN,NACT,   IJDIM,IJTOT,                  94
     2  NVAR,ICARD,ITOT,NVIDI,   Q(NS),Q(ND),Q(NPJ),Q(NCO),Q(NCOOR), 95
     3  Q(NBZ),    Q(NU),Q(NW),Q(NIDJ),Q(NAD),                       96
     5  Q(NV),Q(NX),Q(NY),Q(NID),  NLEG,NSAV,NGUS,NBAND)             97
   50 WRITE (IMP,2400)                                                98
 2000 FORMAT (1H1,52X,19HSTEP   ** CORAN **/1H0,130(1H-))            99
 2100 FORMAT (1H0,26HPARAMETER CARD FOR  CORAN /1H ,26(1H-)/1H0,10X, 100
     1  6HNFAC =,I4,5X,6HLIST3=,I4,5X,6HNGRAF=,I4,5X,6HNPAGE=,I4,    101
     2  5X,6HNLIGN= ,I4,5X,6HJBASE= ,I4,5X,6HNITER= ,I4//)           102
 2200 FORMAT (1H0,42HRE-READING  PARAMETERS ON THE FILES             103
     1 1H ,10X,6HICARD=,I4,5X,6HITOT =,I4,5X,6HNACT =,I4,5X,         104
     2 6HNVAR =,I4,5X,6HNVIDI=,I4)                                   105
 2300 FORMAT (1H0,23HMEMORY USAGE         ,10X,9H YOU HAVE,1X,       106
     1        8HRESERVED,I6,10X,19H  YOU NEED      ,I6/)             107
 2400 FORMAT (1H0/1H0,130(1H-)/1H0,48X,                              108
     1        30HEND OF STEP    ** CORAN  **/1H0,130(1H-))           109
 2500 FORMAT (1H0,21HOUTPUT FILE        = ,I4,3H (,A4,1H) )          110
      RETURN                                                          111
      END                                                             112
                                                                      113
      SUBROUTINE DCALA (NFAC,LIST3,JBASE,LBASE,NITER,NGRAF,NLIGN,NPAGE, 114
     1 KB,   NDIM,NMAX,NQFIN,NACT,  IJDIM,IJTOT,NVAR,                115
     2 ICARD,ITOT,NVIDI,   S,D,PJ,CO,COORD,BZ,U,W,IDJ,AD,V,X,Y,ID,   116
     4               NLEG, NSAV, NGUS, NBAND)                        117
C * * * * * * * * * * * * * * * * * * * * * * * * * * * * * * * * * * 118
C    MAIN SUBROUTINE                                           *     119
C INPUT   NFAC, LIST3,    JBASE, NITER                         *     120
C         NDIM              NVAR = NACT + NSUP                  *     121
C         NMAX              NACT                                *     122
C         NQFIN = NUMBER OF VARIABLES  ON   FILE               *     123
C         NACT  = NUMBER OF ACTIVE COLUMNS                      *     124
C         NVAR  = TOTAL OF  COLUMNS (ACTIVE + ILLUSTRATIVE  )   *     125
C         ICARD = NUMBER  OF ACTIVE ROWS                        *     126
C         ITOT  = TOTAL NUMBER  (ITOT = ICARD + ISUP)           *     127
C         NVIDI =  LENGTH ( A4) OF    IDENTIFIERS OF   ROWS     *     128
```

205

```
C OUTPUT                                                              *    129
C       S(,) , D()                                                    *    130
C       PJ(),CO(),COORD(,) , IDJ()                                    *    131
C    FILES ON INPUT   NLEG                                            *    132
C        ON OUTPUT  NSAV, NGUS   (TEMPORARY FILE NBAND)               *    133
C * * * * * * * * * * * * * * * * * * * * * * * * * * * * * * * * *        134
      DIMENSION S(NMAX,KB),D(NACT),PJ(NVAR),CO(NVAR),AD(LBASE,LBASE),      135
     1 COORD(NDIM,NFAC),U(NQFIN),W(NFAC),IDJ(NQFIN),                       136
     2 TL(18), LT(6), LTITR(20), LIGN(8),BZ(NMAX,LBASE),V(NQFIN),          137
     3 X(IJDIM),Y(IJDIM),ID(IJDIM,NVIDI),IDI(3)                            138
      COMMON / ENSOR / LEC,IMP                                             139
      DATA LT/2HF1,2HF2,2HF3,2HF4,2HF5,2HF6/ ,LBB/4H    /                  140
C......... COPY  OF   DATA ON NSAV, WITH WEIGHTS OF  ROWS                  141
      NQDEB = 0                                                           142
      CALL PRECO (NQFIN,NQDEB,NACT,NVIDI,ITOT,U, NLEG,NSAV)                143
      CALL TITLE (O, NLEG)                                                 144
      READ (NLEG)  (IDJ(J),J=1,NVAR)                                       145
C......... DIAGONALIZATION AND COMPUTATION OF COORDINATES                  146
C        COPY ON  NGUS OF COORDINATES OF ITOT  ROWS                       147
      IF (NFAC .GT. (NACT - 1) )          NFAC = NACT - 1                  148
      IF (JBASE .GT.0 )               CALL LECDI                          149
     1 (JBASE,NACT,NMAX,NQFIN,NVAR,ICARD,NQDEB,NFAC,NITER,                 150
     2 PJ,S,U,TRACE,SOM,V,BZ,D,AD,NSAV,NBAND)                             151
      IF(JBASE.GT.0 )                  GO TO 5                            152
      CALL MEMOI                                                          153
     1 (NMAX,ICARD,NACT,NVAR,NQDEB,NQFIN,S,D,U,PJ,TRACE,SOM,NSAV,NFAC)     154
    5 CONTINUE                                                            155
      JDEB  = 0                                                           156
      CALL AFCOB (NDIM,NMAX,NFAC,NQFIN,NQDEB, JDEB,KB,                     157
     1 NACT,NVAR,NVIDI,ITOT,ICARD,    S,D,TRACE,PJ,SOM,COORD,CO,           158
     2                U,W, NSAV, NGUS )                                    159
C......... PRINTING  OF THE EIGENVALUES                                    160
      WRITE (IMP,1000)                                                    161
      NVP1 = NFAC                                                         162
      IF(JBASE .EQ. 0)  NVP1 = NACT - 1                                   163
      CALL VPOUR (1, NVP1, 15, D, TRACE)                                  164
C......... PRINTING OF COORDINATES AND CONTRIBUTIONS OF  COLUMNS           165
      NF6   = 6                                                          166
      IF (NFAC .LT. 6)                  NF6 = NFAC                        167
      DO 10  L = 1,18                                                     168
   10 TL(L) = 0.0                                                         169
      WRITE (IMP,1100)                                                    170
      WRITE (IMP,1200)                                                    171
      WRITE (IMP,1300) (LT(K),K=1,6),(LT(K),K=1,6),(LT(K),K=1,6)          172
      DO 30  J = 1,NVAR                                                   173
      IF (J .EQ. NACT+1)                WRITE (IMP,1400)                  174
      BB  = 0.0                                                          175
      IF (J .LE. NACT)                  BB = 1                            176
      PJR  = PJ(J) / SOM                                                 177
      DO 20  L = 1,NF6                                                    178
      IF (D(L) .LT. 1.0E-10)            D(L) = 1.0 E-10                   179
      TL(L) = COORD(J,L)                                                 180
      AA   = PJR*TL(L)*TL(L)                                             181
      TL(L+6)= 100.0 * AA*BB / D(L)                                       182
      TL(L+12)= AA / (CO(J)*PJR)                                         183
      IDVAR = IDJ(J+JDEB)                                                 184
   20 CONTINUE                                                            185
      WRITE (IMP,1500) IDVAR,LBB,LBB,PJR,CO(J),(TL(LL),LL=1,18)           186
   30 CONTINUE                                                            187
C......... PRINTING OF COORDINATES AND CONTRIBUTION OF  ROWS               188
      IF ( LIST3 .EQ. 0 )              GO TO 35                          189
      WRITE (IMP,1600)                                                    190
      WRITE (IMP,1200)                                                    191
      WRITE (IMP,1300) (LT(K),K=1,6),(LT(K),K=1,6),(LT(K),K=1,6)          192
   35 CALL OPEN ( 2, NGUS, NFAC, LIGN, KTYP, W )                          193
      IF (NVIDI .GT. 3)                NVIDI = 3                          194
      IDI(2) = LBB                                                       195
      IDI(3) = LBB                                                       196
      DO 50  I = 1,ITOT                                                   197
      READ (NGUS) (W(K),K=1,NFAC), PIA, DOR, (IDI(L),L=1,NVIDI)          198
      IF ( LIST3 .EQ. 0 )              GO TO 50                          199
      IF (I .EQ. ICARD + 1)            WRITE (IMP,1400)                  200
      PIR  = PIA / SOM                                                  201
      BB   = 0.0                                                        202
```

206

```
        IF (I .LE. ICARD)            BB = 1.0                      203
        DO 40  L = 1,NF6                                           204
      TL(L) = W(L)                                                 205
      AA    = PIR * TL(L)*TL(L)                                    206
      TL(L+6)= 100.0 * AA*BB / D(L)                                207
      TL(L+12)= AA / (DOR*PIR)                                     208
   40 CONTINUE                                                     209
      WRITE (IMP,1500) (IDI(L) ,L=1,3), PIR, DOR, (TL(LL),LL=1,18) 210
   50 CONTINUE                                                     211
C........ COPY ON  NGUS, WITHOUT REWIND AFTERWARDS                 212
        DO 60  J = 1,NVAR                                          213
   60 WRITE (NGUS)(COORD(J,K),K=1,NFAC),PJ(J),CO(J),IDJ(JDEB+J),LBB,LBB  214
C.........NGRAF   DISPLAYS                                         215
        IF(NGRAF .EQ.  0 )            RETURN                       216
        IF (NPAGE .GT. 8)            NPAGE = 8                     217
      IX = 0                                                       218
        DO 120 KG = 1,NGRAF                                        219
      REWIND NGUS                                                  220
      IX = IX + 1                                                  221
      IY = IX + 1                                                  222
        IF (IY .GT. NFAC)            RETURN                        223
      CALL OPEN(2,NGUS,NFAC,LIGN,KTYP,W)                           224
        DO 110 I = 1,IJTOT                                         225
      READ( NGUS ) (W(K) ,K=1,NFAC) ,PIA,DOR,(IDI(L),L=1,NVIDI)    226
      X(I) = W(IX)                                                 227
      Y(I) = W(IY)                                                 228
        DO 100  L = 1,NVIDI                                        229
  100 ID(I,L) = IDI(L)                                             230
  110 CONTINUE                                                     231
      CALL HPLAN(IJDIM,IJTOT,IX,IY,X,Y,ID,4,NLIGN,NPAGE,2.5,1,NVIDI,  232
     1          NBAND)                                             233
  120   CONTINUE                                                   234
 1000 FORMAT (1H1)                                                 235
 1100 FORMAT (1H1,10X,40HCOORDINATES    AND    CONTRIBUTIONS        ,  236
     1 14HOF THE COLUMNS/1H ,132(1H-) // )                         237
 1200 FORMAT(1H0,5HNAMES,8X,15HMASSES DIST.  *,12X,12HCOORDINATES ,12X,  238
     1 2H *,4X,9HABSOLUTE ,13HCONTRIBUTIONS,4X,2H *,3X,            239
     2 10H SQUARED  ,13HCORRELATIONS ,4X,2H */ 1H ,130(1H*) )      240
 1300 FORMAT (1H0,26X,2H *,6(3X,A2,1X),2H *,6(2X,A2,1X),           241
     1 2H *,6(2X,A2,1X),2H * /1H ,130(1H*) )                       242
 1400 FORMAT (1H0,24HSUPPLEMENTARY ELEMENTS   / )                  243
 1500 FORMAT (1H ,3A4,1X,F5.3,1X,F6.2,1X,2H *,                     244
     1       6F6.2,2H *,6F5.1,2H *,6F5.2,2H *)                     245
 1600 FORMAT (1H1,10X,40HCOORDINATES    AND    CONTRIBUTIONS        ,  246
     1    11HOF THE ROWS /1H ,132(1H-) // )                        247
      RETURN                                                       248
      END                                                          249
                                                                   250
      SUBROUTINE SELEC (Q, MOTS, NLEG, NBAND)                      251
C * * * * * * * * * * * * * * * * * * * * * * * * * * * * * * * *  252
C      PREPARATION OF FILE NLEG FOR CORRESPONDENCE ANALYSIS    *  253
C CALLS                                                        *  254
C             - LPLUM - CHOIX -  *                             *  255
C                     - PPLUM - LDPLU *                        *  256
C * * * * * * * * * * * * * * * * * * * * * * * * * * * * * * * *  257
      DIMENSION Q(MOTS), LTITR(20)                                 258
      COMMON /ENSOR/ LEC,IMP                                       259
      WRITE(IMP,3000)                                              260
C....... READING OF TITLE                                         261
      READ( LEC , 1000) ( LTITR(L) ,L=1,20)                        262
      WRITE(IMP,4000) (LTITR(L),L=1,20)                            263
C....... READING  OF  PARAMETERS                                  264
      READ(LEC,2000)IEXA, NQEXA, NVIDI, LFMT, MODIG                265
 2000 FORMAT(20I4 )                                                266
      IDIM = IEXA                                                  267
        IF(MODIG .EQ. 0)            IDIM = 1                       268
      WRITE(IMP,5000) IEXA,NQEXA,NVIDI,LFMT,MODIG                  269
        IF(LFMT .EQ.0)            LFMT = 1                         270
      KFMT  = 20*LFMT                                              271
C....... MEMORY RESERVATION                                       272
      NFMT  = 1                                                    273
      NIU   = NFMT + KFMT                                          274
      NIX   = NIU + NQEXA                                          275
      NIG   = NIX + NQEXA                                          276
```

207

```
        NIDJ  = NIG + IDIM                                            277
        NV    = NIDJ + NQEXA                                          278
        NW    = NV + NQEXA                                            279
        NIV   = NW + NQEXA                                            280
        NFIN  = NIV + NQEXA                                           281
        IF ( NFIN .GT. MOTS )          GO TO 50                       282
        CALL LPLUM (MODIG,IDIM,IEXA,NQEXA,NVIDI,KFMT,Q(NFMT),         283
       1          Q(NIU),Q(NIX),Q(NIG),Q(NIDJ),Q(NV),Q(NW),Q(NIV),    284
       2                                   LTITR,NLEG,NBAND)           285
     50 WRITE (IMP,6000)                                              286
   1000 FORMAT (20A4)                                                 287
   3000 FORMAT(1H1,52X,19H STEP   ** SELEC ** ,/1H0,130(1H-))         288
   4000 FORMAT( 1H ,//,5X,7H TITLE=,20A4,/)                           289
   5000 FORMAT(1H ,//,5X,6H IEXA=,I4,5X,6HNQEXA=,I4,5X,6HNVIDI=,I4,   290
       1 5X,6HLFMT =,I4,5X,6HMODIG=,I4,/)                             291
   6000 FORMAT (1H0/1H0,130(1H-)/1H0,48X,                            292
       1          31HEND OF READING AND SELECTION   /1H0,130(1H-))    293
        RETURN                                                        294
        END                                                          295
                                                                      296
        SUBROUTINE LPLUM (MODIG,IDIM,IEXA,NQEXA,NVIDI,KFMT,FMT,       297
       1            IU,IX,IG,IDJ,V,W,IV,LTITR,NLEG,NBAND)             298
C * * * * * * * * * * * * * * * * * * * * * * * * * * * * * * * * * *  299
C 1/ READING  OF IDENTIFIERS OF  COLUMN ( A4)                 *       300
C 2/ CALL  OF CHOIX, READING  OF CODING OF ELEMENTS          *       301
C 3/ READING OF DATA FORMAT  ( LENGTH = KFMT TIMES A4)       *       302
C 4/ CALL  OF PPLUM FOR READING DATA AND CREATING THE FILE NLEG  *   303
C * * * * * * * * * * * * * * * * * * * * * * * * * * * * * * * * * *  304
        DIMENSION LTITR(20),NGR(10),FMT(KFMT)                         305
        DIMENSION IU(NQEXA),IX(NQEXA),IG(IDIM),IDJ(NQEXA),            306
       1            V(NQEXA),W(NQEXA),IV(NQEXA)                       307
        COMMON /ENSOR/ LEC,IMP                                        308
C...... READING  THE IDENTIFIERS  OF  COLUMNS                         309
        WRITE (IMP,2000)                                              310
        READ(LEC,1000) (IDJ(J),J=1,NQEXA)                             311
        WRITE (IMP,3000) (IDJ(J),J=1,NQEXA)                           312
C....... CALLS OF CHOIX FOR   ROWS AND  COLUMNS                       313
        CALL CHOIX (IU, NQEXA)                                        314
        IF (MODIG .NE. 0)        CALL CHOIX (IG,IDIM)                 315
C....... READING THE FORMAT CARD                                      316
        READ(LEC,1000) (FMT(L),L=1,KFMT)                              317
        WRITE (IMP,4000)                                              318
        WRITE (IMP,5000) (FMT(L),L=1,KFMT)                            319
C....... READING  DATA  AND CREATION OF NLEG                          320
        CALL PPLUM      (IDIM,IEXA,ITOT,ICARD,NQEXA,KFMT,             321
       1     NVIDI,IU,IG,IDJ,NGR,V,W,IV,IX,LTITR,FMT, NLEG ,NBAND )   322
   1000 FORMAT (20A4)                                                 323
   2000 FORMAT (1H0,20HNAMES OF COLUMNS          /)                   324
   3000 FORMAT (20(2X,A4))                                            325
   4000 FORMAT (1H0,7HFORMAT   /)                                     326
   5000 FORMAT (1H ,20A4)                                             327
        RETURN                                                        328
        END                                                          329
                                                                      330
        SUBROUTINE PPLUM (IDIM,IEXA,ITOT,ICARD,NQEXA,KFMT,           331
       1     NVIDI,IU,IG,IDJ,NGR,V,W,IV,IX,LTITR,FMT, NLEG ,NBAND )   332
C * * * * * * * * * * * * * * * * * * * * * * * * * * * * * * * * * *  333
C       PERMUTATION  OF NAMES AND  OF DATA                    *       334
C       CREATION OF FILE NLEG. CALL ... LDPLU.                *       335
C * * * * * * * * * * * * * * * * * * * * * * * * * * * * * * * * * *  336
        DIMENSION IU(NQEXA),NGR(10),V(NQEXA),W(NQEXA),IV(NQEXA),IDJ(NQEXA) 337
        DIMENSION IX(NQEXA),NGCUM(11),LQ(11),IDENT(15), FMT(KFMT),    338
       1          IG(IDIM),LTITR(20)                                  339
        COMMON  / ENSOR / LEC,IMP                                     340
        DATA  LLEG/4HNLEG/                                            341
        NCOD  = 10                                                    342
        DO 10 KA = 1,NCOD                                             343
        LQ(KA)= 0                                                     344
     10 NGR(KA)= 0                                                    345
        DO 15 K = 1,NQEXA                                             346
        IF (IU(K) .EQ. 0)             IU(K) = NCOD                    347
        KA   = IU(K)                                                  348
     15 NGR(KA)= NGR(KA) + 1                                          349
        NGCUM(1)= 0                                                   350
```

```
          DO 20  KA = 1,NCOD                                            351
   20 NGCUM(KA+1)= NGCUM(KA) + NGR(KA)                                  352
C ........ PERMUTATION VECTOR OF DATA  IX(*)                            353
          DO 25  K = 1,NQEXA                                            354
      KA   = IU(K)                                                      355
      LQ(KA)= LQ(KA) + 1                                                356
   25 IX(K) = LQ(KA) + NGCUM(KA)                                        357
C ........ PERMUTATION OF IDJ                                           358
          DO 30  J = 1,NQEXA                                            359
   30 IV(J) = IDJ(J)                                                    360
          DO 40 K = 1,NQEXA                                             361
      KD = IX(K)                                                        362
   40 IDJ(KD) = IV(K)                                                   363
C ........ PERMUTATION OF  DATA  AND PRINTING ON  NLEG                  364
      ICARD = 0                                                         365
      ISUP  = 0                                                         366
      REWIND NLEG                                                       367
      REWIND NBAND                                                      368
      JCARD = NGR(1)                                                    369
      JTOT = NGR(1) + NGR(2)                                            370
      IF( IDIM .LE.1)    GO TO 60                                       371
      DO 50 I = 1,IEXA                                                  372
      IF(IG(I) .EQ. 1 )              ICARD = ICARD + 1                  373
      IF(IG(I) .EQ. 2 )              ISUP  = ISUP + 1                   374
   50 CONTINUE                                                          375
                                     GO TO 70                           376
   60 ICARD = IEXA                                                      377
      ITOT = ICARD                                                      378
   70 CONTINUE                                                          379
      ITOT = ICARD + ISUP                                               380
C........ CREATION OF FILE NLEG ...FOR CORAN                            381
      WRITE(NLEG) (LTITR(L),L=1,20),LLEG,ICARD,ITOT,JCARD,JTOT,NVIDI    382
      WRITE(NLEG) (IDJ(J),J=1,JTOT)                                     383
      POIDS = 1.0                                                       384
      DO 150 I = 1,IEXA                                                 385
      CALL LDPLU (LEC, NVIDI, NQEXA, IDENT, W, FMT,KFMT)                386
      DO 130  K = 1,NQEXA                                               387
      KD   = IX(K)                                                      388
  130 V(KD) = W(K)                                                      389
      IF (IDIM .LE. 1)              GO TO 140                           390
      IF (IG(I).EQ. 0)              GO TO 150                           391
      IF (IG(I).EQ. 2)              GO TO 145                           392
  140 WRITE (NLEG)  (V(KK), KK=1,JTOT),POIDS,(IDENT(L),L=1,NVIDI)       393
                                    GO TO 150                           394
  145 WRITE (NBAND) (V(KK), KK=1,JTOT),POIDS,(IDENT(L),L=1,NVIDI)       395
  150 CONTINUE                                                          396
C.......... COPY OF SUPPLEMENTARY INDIVIDUALS (OR ROWS)                 397
      IF (ISUP .EQ. 0)              GO TO 170                           398
      REWIND NBAND                                                      399
      DO 160  I = 1,ISUP                                                400
      READ (NBAND)  (V(KK), KK=1,JTOT),POIDS,(IDENT(L),L=1,NVIDI)       401
  160 WRITE (NLEG)  (V(KK), KK=1,JTOT),POIDS,(IDENT(L),L=1,NVIDI)       402
  170 CONTINUE                                                          403
C ....... RECONSTRUCTION OF IU(*)                                       404
          DO 180  K = 1,NQEXA                                           405
  180 IF (IU(K) .EQ. NCOD)              IU(K) = 0                       406
      RETURN                                                            407
      END                                                               408
                                                                        409
      SUBROUTINE GSMOD  (IDIM,ICARD,JCARD,P,X,KRANG,T,V)                410
C * * * * * * * * * * * * * * * * * * * * * * * * * * * * * * * *       411
C     ORTHONORMALIZATION OF JCARD (FIRST) COLUMNS OF X(ICARD,*)  *      412
C     BY MODIFIED GRAM-SCHMIDT METHOD                            *      413
C * * * * * * * * * * * * * * * * * * * * * * * * * * * * * * * *       414
      DIMENSION  X(IDIM,JCARD) , P(IDIM) , T(IDIM) , V(IDIM)            415
      DATA EPS / 1.0 E-10 /                                             416
      KRANG = JCARD                                                     417
      DO 20  J = 1,JCARD                                                418
      V(J)   = 0.0                                                      419
      DO 10  I = 1,ICARD                                                420
   10 V(J)   = V(J) + P(I)*X(I,J)*X(I,J)                                421
      IF (V(J) .LE. 1.E-10)         V(J) = 1.0 E-10                     422
   20 CONTINUE                                                          423
      C    = 1.0 / SQRT(V(1))                                           424
```

209

```
        DO 30  I = 1,ICARD                                              425
 30 X(I,1) = C*X(I,1)                                                   426
       IF (JCARD .EQ. 1)              GO TO  130                        427
     KFIN   = JCARD - 1                                                 428
        DO 120  J = 1,KFIN                                              429
     J1     = J + 1                                                     430
        DO 60  JJ = J1,JCARD                                            431
     T(JJ) = 0.0                                                        432
        DO 40  I = 1,ICARD                                             433
 40 T(JJ)  = T(JJ) + P(I)*X(I,JJ)*X(I,J)                                434
        DO 50  I = 1,ICARD                                              435
 50 X(I,JJ)= X(I,JJ) - T(JJ)*X(I,J)                                     436
     60   CONTINUE                                                      437
C....... TEST OF COLLINEARITY. NORMALIZATION OF X(*,J1).               438
     C      = 0.0                                                       439
        DO 70  I = 1,ICARD                                              440
 70 C      = C + P(I)*X(I,J1)*X(I,J1)                                   441
       IF (C/V(J1) - EPS)            80 , 80 , 90                       442
 80 C      = 0.0                                                        443
     KRANG  = KRANG - 1                                                 444
                                     GO TO 100                          445
 90 C      = 1.0 / SQRT(C)                                              446
100    DO 110  I = 1,ICARD                                              447
110 X(I,J1)= C*X(I,J1)                                                  448
120    CONTINUE                                                         449
130    CONTINUE                                                         450
       RETURN                                                           451
       END                                                              452
                                                                        453
     SUBROUTINE   TRIDI ( NDIM, N, W, D, S )                            454
C * * * * * * * * * * * * * * * * * * * * * * * * * * * * * * * * * *   455
C HOUSEHOLDER REDUCTION TO TRIDIAGONAL FORM                         *   456
C REFERENCES                                                        *   457
C     1/ J.H.WILKINSON-C.REINSCH / HANDBOOK FOR AUTOMATIC COMPUTATION*  458
C        VOLUME 2, SPRINGER-VERLAG, 1971 /                          *   459
C     2/ MATRIX EIGENSYSTEM ROUTINES, EISPAK GUIDE / LECTURE NOTES IN* 460
C        COMPUTER SCIENCE NO.6 / SPRINGER-VERLAG , 1974 /           *   461
C * * * * * * * * * * * * * * * * * * * * * * * * * * * * * * * * * *   462
     DIMENSION  D(N) , S(N) , W(NDIM,N)                                 463
       IF (N .EQ. 1)        GO TO 130                                   464
        DO 120  I2 = 2,N                                                465
     B      = 0.0                                                       466
     C      = 0.0                                                       467
     I      = N - I2 + 2                                                468
     K      = I - 1                                                     469
       IF (K .LT. 2)        GO TO  20                                   470
        DO 10  L=1,K                                                    471
 10 C      = C + ABS(W(I,L))                                            472
       IF (C .NE. 0.0)      GO TO  30                                   473
 20 S(I)   = W(I,K)                                                     474
                            GO TO 110                                   475
 30    DO 40  L= 1,K                                                    476
     W(I,L)= W(I,L) / C                                                 477
     B      = B + W(I,L)*W(I,L)                                         478
 40    CONTINUE                                                         479
     P      = W(I,K)                                                    480
     Q      = -SIGN (SQRT(B) , P)                                       481
     S(I)   = C * Q                                                     482
     B      = B - P*Q                                                   483
     W(I,K)= P - Q                                                      484
     P      = 0.0                                                       485
        DO 80   M = 1,K                                                 486
     W(M,I)= W(I,M) / (B*C)                                             487
     Q      = 0.0                                                       488
        DO 50  L = 1,M                                                  489
 50 Q      = Q + W(M,L)*W(I,L)                                          490
     M1     = M + 1                                                     491
       IF (K .LT. M1)        GO TO  70                                  492
        DO 60  L = M1,K                                                 493
 60 Q      = Q + W(L,M)*W(I,L)                                          494
 70 S(M)   = Q / B                                                      495
     P      = P + S(M)*W(I,M)                                           496
 80    CONTINUE                                                         497
     PB     = P / (B+B)                                                 498
```

210

```
      DO   90   M = 1,K                                          499
      P     = W(I,M)                                             500
      Q     = S(M) - PB*P                                        501
      S(M)  = Q                                                  502
      DO   90   L = 1,M                                          503
      W(M,L)= W(M,L) - P*S(L) - Q*W(I,L)                         504
   90 CONTINUE                                                   505
      DO 100   L = 1,K                                           506
  100 W(I,L)= C * W(I,L)                                         507
  110 D(I)  = B                                                  508
  120 CONTINUE                                                   509
  130 S(1)  = 0.0                                                510
      D(1)  = 0.0                                                511
      DO 180   I = 1,N                                           512
      K     = I - 1                                              513
      IF (D(I) .EQ. 0.0)   GO TO  160                            514
      DO 150   M = 1,K                                           515
      Q     = 0.0                                                516
      DO 140   L = 1,K                                           517
  140 Q     = Q + W(I,L)*W(L,M)                                  518
      DO 150   L = 1,K                                           519
      W(L,M)= W(L,M) - Q*W(L,I)                                  520
  150 CONTINUE                                                   521
  160 D(I)  = W(I,I)                                             522
      W(I,I)= 1.0                                                523
      IF (K .LT. 1)        GO TO  180                            524
      DO 170   M = 1,K                                           525
      W(I,M)= 0.0                                                526
      W(M,I)= 0.0                                                527
  170 CONTINUE                                                   528
  180 CONTINUE                                                   529
      RETURN                                                     530
      END                                                        531
                                                                 532
      SUBROUTINE  VPOUR ( MODE, JCARD, JEDIT, D, TRACE )         533
C * * * * * * * * * * * * * * * * * * * * * * * * * * * * * *    534
C       PRINTING THE EIGENVALUES   D(I),I=1,JCARD           *    535
C       HISTOGRAM  OF FIRST JEDIT VALUES                    *    536
C * * * * * * * * * * * * * * * * * * * * * * * * * * * * * *    537
      DIMENSION  D(JCARD)                                        538
      COMMON / ENSOR / LEC,IMP                                   539
      DATA   IAST / 1H* /                                        540
      WRITE (IMP,100)                                            541
      JFIN  = JEDIT + MODE - 1                                   542
      IF (JFIN .GT. JCARD)      JFIN = JCARD                     543
      IF (MODE .EQ. 1)          GO TO  10                        544
      WRITE (IMP,110)   D(1)                                     545
   10 CONTINUE                                                   546
      WRITE (IMP,120)  TRACE                                     547
      IF (JEDIT .LE. 0)         GO TO  30                        548
      WRITE (IMP,130)                                            549
      PCUM  = 0.0                                                550
      DO 20   J = MODE,JFIN                                      551
      PRC   = 100.0*D(J)/TRACE                                   552
      PCUM  = PCUM + PRC                                         553
      NAST  = 1.0 + 70.0*D(J)/D(MODE)                            554
      IF (NAST .LT. 1)          NAST = 1                         555
      J1    = J - MODE + 1                                       556
      WRITE (IMP,140)  J1,D(J),PRC,PCUM,(IAST,N=1,NAST)          557
   20 CONTINUE                                                   558
   30 JDEB  = JFIN + 1                                           559
      IF (JDEB .GT. JCARD)         RETURN                        560
      J1    = JFIN - MODE + 2                                    561
      J2    = JCARD - MODE + 1                                   562
      WRITE (IMP,150)  J1 , J2                                   563
      WRITE (IMP,160)  (D(J),J=JDEB,JCARD)                       564
  100 FORMAT(//1H ,10X,27HEIGENVALUES                 /1H ,130(1H-)//)  565
  110 FORMAT (1H ,40HTHE FIRST EIGENVALUE(ARTEFACT) IS        ,  566
     1            9H REMOVED  ,F15.8 // )                        567
  120 FORMAT (1H ,25HSUM OF THE EIGENVALUES   ,16X,F15.8 // )    568
  130 FORMAT (1H ,41HHISTOGRAM OF THE FIRST EIGENVALUES     //1H ,5X,  569
     1 13HEIGENVALUE   , 2(13H   PERCENTAGE)/1H ,37X,6HCUM. ,/ )  570
  140 FORMAT (1H ,1X,I3,F15.8,2(4X,F6.2,4X),2X,80A1 )            571
  150 FORMAT (///1H ,39HABBREVIATED SUMMARY OF EIGENVALUES FROM,  572
```

211

```
    1                          I4,3H TO, I4 /  )                          573
  160 FORMAT (1H ,10F12.8 )                                               574
        RETURN                                                            575
        END                                                               576
                                                                          577
        SUBROUTINE  VPROP ( NDIM, N, W, D, S, KODE )                      578
C * * * * * * * * * * * * * * * * * * * * * * * * * * * * * * * * * * *   579
C COMPUTATION OF EIGENVALUES AND EIGENVECTORS                        *    580
C REFERENCES                                                         *    581
C     1/ J.H.WILKINSON , C.REINSCH / HANDBOOK FOR AUTOMATIC COMPUTAT.*    582
C        VOLUME 2 / SPRINGER-VERLAG , 1971 /                        *     583
C     2/ MATRIX EIGENSYSTEM ROUTINES , EISPAK GUIDE /LECTURE NOTES IN*    584
C        COMPUTER SCIENCE NO.6 / SPRINGER-VERLAG , 1974 /           *     585
C * * * * * * * * * * * * * * * * * * * * * * * * * * * * * * * * * * *   586
        DIMENSION  W(NDIM,N) , D(N) , S(N)                                587
        DATA SEUIL / 1.0E-7 /                                             588
        KODE  = 0                                                         589
        CALL  TRIDI (NDIM,N,W,D,S)                                        590
        IF (N .EQ. 1)        GO TO 140                                    591
        DO 10  I = 2,N                                                    592
   10 S(I-1)= S(I)                                                        593
        S(N)  = 0.0                                                       594
        DO 90  K = 1,N                                                    595
        M     = 0                                                         596
   20   DO 30  J = K,N                                                    597
        IF (J .EQ. N)        GO TO  40                                    598
        ABJ   = ABS(S(J))                                                 599
        EPS   = SEUIL*(ABS(D(J)) + ABS(D(J+1)))                           600
        IF (ABJ .LE. EPS)    GO TO  40                                    601
   30   CONTINUE                                                          602
   40 H     = D(K)                                                        603
        IF (J .EQ. K)        GO TO  90                                    604
        IF (M .EQ. 30)       GO TO 130                                    605
        M     = M + 1                                                     606
        Q     = (D(K+1) - H) / (2.0*S(K))                                 607
        T     = SQRT (Q*Q + 1.0)                                         608
        Q     = D(J) - H + S(K) / (Q + SIGN(T,Q))                         609
        U     = 1.0                                                       610
        V     = 1.0                                                       611
        H     = 0.0                                                       612
        JK    = J - K                                                     613
        DO 80 IJK = 1,JK                                                  614
        I     = J - IJK                                                   615
        P     = U * S(I)                                                  616
        B     = V * S(I)                                                  617
        IF (ABS(P).LT.ABS(Q))GO TO   50                                   618
        V     = Q / P                                                     619
        T     = SQRT (V*V + 1.0)                                         620
        S(I+1)= P * T                                                     621
        U     = 1.0 / T                                                   622
        V     = V * U                                                     623
                        GO TO   60                                        624
   50 U     = P / Q                                                       625
        T     = SQRT (U*U + 1.0)                                         626
        S(I+1)= Q * T                                                     627
        V     = 1.0 / T                                                   628
        U     = U * V                                                     629
   60 Q     = D(I+1) - H                                                  630
        T     = (D(I) - Q)*U + 2.0*V*B                                    631
        H     = U * T                                                     632
        D(I+1)= Q + H                                                     633
        Q     = V*T - B                                                   634
        DO 70  L = 1,N                                                    635
        P     = W(L,I+1)                                                  636
        W(L,I+1)= U*W(L,I) + V*P                                          637
        W(L,I)= V*W(L,I) - U*P                                            638
   70   CONTINUE                                                          639
   80   CONTINUE                                                          640
        D(K)  = D(K) - H                                                  641
        S(K)  = Q                                                         642
        S(J)  = 0.0                                                       643
                        GO TO   20                                        644
   90   CONTINUE                                                          645
        DO 120  IJ = 2,N                                                  646
```

212

```
      I    = IJ - 1                                                    647
      L    = I                                                         648
      H    = D(I)                                                      649
         DO 100 M = IJ,N                                               650
         IF (D(M) .LE. H)      GO TO  100                              651
      L    = M                                                         652
      H    = D(M)                                                      653
 100  CONTINUE                                                         654
         IF (L .EQ. I)         GO TO  120                              655
      D(L)  = D(I)                                                     656
      D(I)  = H                                                        657
         DO 110  M = 1,N                                               658
      H    = W(M,I)                                                    659
      W(M,I)= W(M,L)                                                   660
      W(M,L)= H                                                        661
 110    CONTINUE                                                       662
 120    CONTINUE                                                       663
                              GO TO  140                               664
 130 KODE  = K                                                         665
 140    RETURN                                                         666
         END                                                           667
                                                                      668
      SUBROUTINE AFCOB (NDIM,NMAX,NFAC,NQFIN,NQDEB, JDEB,KB,           669
     1 NACT,NVAR,NVIDI,ITOT,ICARD,  S,D,TRACE, PJ,SOM,                 670
     2 COORD,CO,  U, W,   NSAV, NGUS)                                  671
C * * * * * * * * * * * * * * * * * * * * * * * * * * * * * **         672
C CALCULATION OF MATRIX TO BE DIAGONALISED                   *         673
C INPUT   NDIM  = GREATER THAN NVAR = NACT + NSUP            *         674
C         NMAX  = GREATER THAN NACT                          *         675
C         KB    = NACT OR JBASE IF JBASE .GT. 0              *         676
C         NFAC  = NUMBER OF FACTORS                          *         677
C         NQFIN=NUMBER OF VARIABLES RECORDED (NACT+NSUP)     *         678
C         NACT  = NUMBER OF  COLUMNS(ACTIVE)                 *         679
C         NVAR  = NUMBER(TOTAL)OF  COLUMNS (=NACT + NSUP)    *         680
C         NVIDI =   LENGTH (IN A4)  OF    IDENTIFIERS OF  ROW *        681
C         ITOT  = NUMBER TOTAL OF   ROWS (= ICARD + ISUP)    *         682
C         ICARD = NUMBER OF    ROWS(ACTIVE).                 *         683
C OUTPUT       CO(NVAR) = DISTANCES  OF  COLUMNS TO ORIGIN   *         684
C              COORD(NDIM,NFAC)= COORDINATES OF  COLUMNS     *         685
C    FILES ... NSAV IN READING, NGUS IN PRINTING.            *         686
C * * * * * * * * * * * * * * * * * * * * * * * * * * * * * **         687
      DIMENSION S(NMAX,KB),D(NACT),CO(NVAR),COORD(NDIM,NFAC),          688
     1 PJ(NVAR),U(NQFIN),W(NFAC),LTITR(20),IDENT(15),LIGN(8)           689
      DATA LGUS /4HNGUS/                                               690
      ISUP  = ITOT - ICARD                                             691
      NSUP  = NVAR - NACT                                              692
      KTYP  = 2                                                        693
      JACT  = 0                                                        694
         IF (NFAC .GT. NACT-1)            NFAC = NACT - 1              695
      NBID  = 0                                                        696
      REWIND  NSAV                                                     697
      READ (NSAV)  (LTITR(L),L=1,20)                                   698
      REWIND NGUS                                                      699
      WRITE (NGUS)  (LTITR(L),L=1,20), LGUS                            700
      WRITE (NGUS) NFAC,ICARD,ISUP,JACT,JDEB,NACT,NSUP,NBID,NBID,KTYP  701
      WRITE (NGUS)  (D(L), L=1,NFAC), TRACE                            702
         DO 130  J = 1,NVAR                                            703
      CO(J) = 0.0                                                      704
 130 PJ(J) = PJ(J) / SOM                                              705
      NIDET = NVIDI                                                    706
         IF (NIDET .LE. 0)                NIDET = 1                    707
         DO 180  I = 1,ITOT                                            708
      READ (NSAV)  (U(J),J=1,NQFIN), PIA, (IDENT(L),L=1,NIDET)         709
         DO 150  K = 1,NFAC                                            710
      W(K)  = 0.0                                                      711
         DO 140  J = 1,NACT                                            712
 140 W(K)  = W(K) + (U(NQDEB+J)*S(J,K)) / (PIA)                       713
 150    CONTINUE                                                       714
      DOR   = 0.0                                                      715
         DO 160  J = 1,NACT                                            716
 160 DOR   = DOR + (U(NQDEB+J)/PIA - PJ(J))**2 / PJ(J)                 717
         IF (DOR .LT. 1.0E-10)            DOR = 1.0 E-10               718
      PI    = PIA / SOM                                                719
         IF (I .GT. ICARD)               GO TO  175                    720
```

213

```
         DO 170  J = 1,NVAR                                              721
170 CO(J) = CO(J) + (U(NQDEB+J)/(SOM*PJ(J)) - PI)**2 / PI               722
175 WRITE (NGUS)  (W(K),K=1,NFAC), PIA, DOR, (IDENT(L),L=1,NIDET)        723
180     CONTINUE                                                        724
         DO 200  K = 1,NFAC                                             725
         IF (D(K) .LT. 1.0E-9)             D(K) = 1.0 E-9               726
         D(K) = SQRT(D(K))                                             727
         DO 190  J = 1,NACT                                            728
190 COORD(J,K)=S(J,K) * D(K)                                           729
200     CONTINUE                                                       730
C.......... ILLUSTRATIVE COLUMNS                                       731
         DO 210  J = 1,NVAR                                            732
210 PJ(J) = SOM*PJ(J)                                                  733
         IF (NSUP .LE. 0)                  GO TO  260                  734
         J1    = NACT + 1                                              735
         DO 230  JS = J1,NVAR                                          736
         DO 220  K  = 1 ,NFAC                                          737
220 COORD(JS,K)= 0.0                                                   738
230     CONTINUE                                                       739
         CALL TITLE ( 0, NSAV)                                         740
         CALL OPEN ( 2, NGUS, NFAC, LIGN, KTYP, W)                     741
         DO 250  I = 1,ICARD                                           742
         READ (NSAV) (U(J),J=1,NQFIN)                                  743
         READ (NGUS) (W(K),K=1,NFAC)                                   744
         DO 240  JS = J1,NVAR                                          745
         DO 240  K  = 1 ,NFAC                                          746
240 COORD(JS,K)= COORD(JS,K) + (U(NQDEB+JS)*W(K))/                     747
    1                         (PJ(JS)*D(K))                            748
250     CONTINUE                                                       749
260     CONTINUE                                                       750
         DO 270  K = 1,NFAC                                            751
270 D(K)  = D(K)*D(K)                                                  752
         RETURN                                                        753
         END                                                           754
                                                                       755
         SUBROUTINE CPROJ( ICARD , NMAX , JBASE , NACT , NQFIN ,       756
    1 NQDEB , S , D , BB , U , AD , SOMP , PJ,NB )                     757
C * * * * * * * * * * * * * * * * * * * * * * * * * * * * * * * * *     758
C OPERATION OF PROJECTION AND DIAGONALIZATION, FOR DIAGONALIZATION  *  759
C BY DIRECT READING                                                *  760
C * * * * * * * * * * * * * * * * * * * * * * * * * * * * * * * * *     761
         DIMENSION S(NMAX,JBASE) , BB(NMAX,JBASE) , AD(JBASE,JBASE) ,  762
    1 U(NQFIN),PJ(NACT) , D(NACT)                                     763
         CALL TITLE(0,NB)                                              764
         DO 10 L = 1,JBASE                                             765
         DO 10 M = 1,JBASE                                             766
10 AD(L,M) = 0.0                                                       767
         DO 40 I = 1,ICARD                                             768
         READ( NB) (U(KK),KK=1,NQFIN) ,PIA                             769
         PQ = 1. / (SOMP*PIA)                                          770
         DO 30 M = 1,JBASE                                             771
         DO 30 L = 1,M                                                 772
         CIL = 0                                                       773
         CIM = 0                                                       774
         DO 20 K = 1,NACT                                              775
         UUU = U(K+NQDEB)  -PJ(K)*PIA                                  776
         CIL = CIL + S(K,L) * UUU                                      777
20 CIM = CIM + S(K,M) * UUU                                            778
30 AD(L,M) = AD(L,M) + PQ *CIL *CIM                                    779
40 CONTINUE                                                            780
         DO 50 M = 1,JBASE                                             781
         DO 50 L = 1,M                                                 782
50 AD(M,L) = AD(L,M)                                                   783
C.......DIAGONALIZATION OF THE PROJECTED MATRIX....                    784
         CALL VPROP( JBASE , JBASE , AD , D , U , KOD)                 785
         DO 70 M = 1,JBASE                                             786
         DO 70 J = 1,NACT                                              787
         BB(J,M) = 0                                                   788
         DO 70 K = 1,JBASE                                             789
         BB(J,M) = BB(J,M) + S(J,K) * AD(K,M)                          790
70 CONTINUE                                                            791
         DO 80 J = 1,NACT                                              792
         DO 80 L = 1,JBASE                                             793
80 S(J,L) = BB(J,L)                                                    794
```

214

```
      RETURN                                                           795
      END                                                              796
                                                                       797
      SUBROUTINE LECDI( JBASE,NACT,NMAX,NQFIN,NVAR,ICARD,              798
     1 NQDEB,KFAC,NITER,PJ,S,U,TRACE,SOMP,V,BB,D,AD,NB,NBAND)          799
C * * * * * * * * * * * * * * * * * * * * * * * * * * * * * * * * *    800
C   ANALYSIS OF CORRESPONDENCES .DIAGONALIZATION  BY DIRECT READING  * 801
C   PROGRAM  CALLED  ... TITLE, GSMOD, PUISC, CPROJ.               *   802
C * * * * * * * * * * * * * * * * * * * * * * * * * * * * * * * * *    803
      DIMENSION LTITR(10) , S(NMAX,JBASE) , PJ(NVAR) , U(NQFIN)        804
     1 , BB(NMAX,JBASE) , V(NQFIN) , D(NACT) , AD(JBASE,JBASE)         805
      COMMON / ENSOR / LEC,IMP                                         806
      NAR = 10                                                         807
      NZERO = NITER - 1                                                808
      TRACE = 0                                                        809
      DO 10 J = 1,NVAR                                                 810
      V(J) = 0                                                         811
   10 PJ(J) = 0                                                        812
      CALL TITLE(0,NB)                                                 813
      SOMP = 0.0                                                       814
C    CALCULATION OF WEIGHTS                                            815
      DO 40 I = 1,ICARD                                                816
      READ(NB) ( U(J) ,J = 1,NQFIN) , PIA                             817
      SOMP =  SOMP + PIA                                               818
      DO 20  J = 1,NVAR                                                819
   20 PJ(J) = PJ(J) + U(NQDEB + J)                                     820
      DO 30 J = 1,NACT                                                 821
   30 V(J) = V(J) + U(NQDEB + J)**2 /PIA                               822
   40 CONTINUE                                                         823
      DO 45 J = 1,NACT                                                 824
   45 TRACE = TRACE + V(J)/PJ(J)                                       825
      TRACE = TRACE - 1                                                826
      WRITE(IMP,640) TRACE                                             827
      DO 46 J = 1,NACT                                                 828
   46 PJ(J) = PJ(J)/SOMP                                               829
      DO 50 J = 1,NACT                                                 830
      DO 50 L = 1,JBASE                                                831
   50 S(J,L) = SEN3A(BID)                                              832
      CALL GSMOD (NMAX,NACT,JBASE,PJ,S,KR,U,V)                         833
      DO 80 NIT = 1,NITER                                              834
      REWIND NBAND                                                     835
      DO 60 L = 1,JBASE                                                836
   60 WRITE(NBAND) (S(J,L),J=1,NACT)                                   837
      CALL PUISC( ICARD,NMAX,JBASE,NQDEB,NACT,NQFIN,NAR,PJ,S,BB,       838
     1 U,V,SOMP,TRACE,NB)                                              839
      IF(NIT.LT.NZERO)      GO TO 80                                   840
      CALL CPROJ (ICARD,NMAX,JBASE,NACT,NQFIN,NQDEB,S,D,BB,U,AD,SOMP,  841
     1 PJ,NB )                                                         842
      WRITE(IMP,610)  NIT                                              843
      WRITE( IMP , 630 )( D(J) ,J=1,KFAC )                             844
      IF (NIT .EQ. NZERO)      GO TO 80                                845
      REWIND NBAND                                                     846
      DO 70 L = 1,KFAC                                                 847
      V(L) = 0.0                                                       848
      READ(NBAND) (U(J),J=1,NACT)                                      849
      DO 70 J = 1,NACT                                                 850
   70 V(L) = V(L) + PJ(J)*S(J,L)*U(J)                                  851
      WRITE(IMP,620)                                                   852
      WRITE(IMP,630) (V(L),L=1,KFAC)                                   853
   80 CONTINUE                                                         854
      DO 90 J=1,NACT                                                   855
   90 PJ(J) = PJ(J)*SOMP                                               856
  610 FORMAT (//1H ,10X,10HREAD         ,I3, / 1H ,132(1H-)//1H ,10X,  857
     1   47HSUMMARY OF PRIMARY EIGENVALUES              / )            858
  620 FORMAT (//1H ,10X,40HCOSINES OF EIGENVECTORS             / )     859
  630 FORMAT (1H ,10F12.8)                                             860
  640 FORMAT (1H0,10X,7HTRACE =,F12.8/)                                861
      RETURN                                                           862
      END                                                              863
                                                                       864
```

```
      SUBROUTINE MEMOI                                                   865
     1 (NMAX,ICARD,NACT,NVAR,NQDEB,NQFIN,S,D,U,PJ,TRACE,SOM,NSAV,NFAC)    866
C * * * * * * * * * * * * * * * * * * * * * * * * * * * * * * * * *       867
C    ANALYSIS OF CORRESPONDENCE .COMPUTATION OF THE MATRIX TO BE    *     868
C    DIAGONALIZED  (IN CENTRAL MEMORY)                              *     869
C * * * * * * * * * * * * * * * * * * * * * * * * * * * * * * * * *       870
      DIMENSION S(NMAX,NACT),D(NACT),U(NQFIN),LTITR(20),PJ(NVAR)          871
      COMMON/ ENSOR / LEC,IMP                                             872
      DO 10  J = 1,NVAR                                                   873
 10 PJ(J)= 0.0                                                           874
      DO 30  J = 1,NACT                                                   875
      DO 20  JJ = 1,J                                                     876
 20 S(J,JJ)= 0.0                                                         877
 30   CONTINUE                                                           878
      REWIND NSAV                                                         879
      READ (NSAV)  (LTITR(LL), LL=1,20)                                   880
      SOM   = 0.0                                                         881
      DO 70  I = 1,ICARD                                                  882
      READ (NSAV)  (U(J),J=1,NQFIN), PIA                                  883
      IF (PIA .LE. 1.0 E-8)        GO TO  70                              884
      SOM   = SOM + PIA                                                   885
      DO 50  J = 1,NACT                                                   886
      DO 40  JJ = 1,J                                                     887
 40 S(J,JJ)= S(J,JJ) + (U(NQDEB+J)*U(NQDEB+JJ))/PIA                       888
 50   CONTINUE                                                           889
      DO 60  J = 1,NVAR                                                   890
 60 PJ(J) = PJ(J) + U(NQDEB+J)                                           891
 70   CONTINUE                                                           892
      DO 80  J = 1,NVAR                                                   893
      IF (PJ(J) .LT. 1.E-9)        PJ(J) = 1.0 E-9                        894
 80   CONTINUE                                                           895
      DO 100  J = 1,NACT                                                  896
      DO 90  JJ = 1,J                                                     897
      BBB   = PJ(J)*PJ(JJ)                                                898
      AAAA  = SQRT(BBB)                                                   899
      S(J,JJ) = ( S(J,JJ) / AAAA ) - ( AAAA / SOM )                       900
 90 S(JJ,J)= S(J,JJ)                                                     901
 100  CONTINUE                                                           902
      TRACE = 0.0                                                         903
      DO 110  J = 1,NACT                                                  904
 110 TRACE = TRACE + S(J,J)                                              905
C.......... DIAGONALIZATION                                              906
      CALL VPROP (NMAX, NACT, S, D, U, KODE)                              907
      IF (KODE .EQ. 0)              GO TO  120                            908
      WRITE (IMP,1000)  KODE                                             909
 1000 FORMAT (1H0/1H0,10X,37HFATAL ERROR  IN      DIAGONALISATION/        910
     1        1H0,10X,6HKODE =,I4 / 1H0,132(1H-) // )                     911
      STOP                                                                912
 120 CONTINUE                                                            913
      DO 140  L = 1,NFAC                                                  914
      DO 130  K = 1,NACT                                                  915
      IF (PJ(K) .LT. 1.0E-8)        PJ(K) = 1.0E-8                        916
 130 S(K,L)= S(K,L)*SQRT(SOM/PJ(K))                                      917
 140  CONTINUE                                                           918
      RETURN                                                              919
      END                                                                920
                                                                         921
      SUBROUTINE PRECO (NQFIN,NQDEB,NACT,NVIDI,ITOT, U, NLEG,NSAV)        922
C * * * * * * * * * * * * * * * * * * * * * * * * * * * * * * * * * *     923
C PREPARATION FOR CORRESPONDENCE ANALYSIS                          *      924
C THE DATA READ FROM NLEG IS SAVED ON NSAV WITH A COPY OF THE      *      925
C ABSOLUTE WEIGHTS(SUM OF ACTIVE FREQUENCIES)                      *      926
C INPUT    NQFIN = NUMBER  OF VARIABLES RECORDED                   *      927
C          NACT = NUMBER OF  ACTIVE COLUMNS                        *      928
C THERE ARE NQFIN - NACT ILLUSTRATIVE COLUMNS                      *      929
C          NVIDI =  LENGTH (IN A4) OF   IDENTIFIERS OF   ROW       *      930
C          ITOT = TOTAL NUMBER OF ROWS                             *      931
C * * * * * * * * * * * * * * * * * * * * * * * * * * * * * * * * * *     932
      DIMENSION U(NQFIN), LTITR(20), IDENT(15)                            933
      DATA LSAV /4HNSAV/ , LLEG/4HNLEG/                                   934
      REWIND NLEG                                                         935
      REWIND NSAV                                                         936
      READ (NLEG)  (LTITR(L),L=1,20), NXX                                 937
      IF (NXX .EQ. LLEG)            READ (NLEG)  NBID                     938
      WRITE(NSAV)  (LTITR(L),L=1,20), LSAV                                939
      NQD   = NQDEB + 1                                                   940
      NIDET = NVIDI                                                       941
      IF (NIDET .LE. 0)                 NIDET = 1                         942
```

216

```
          DO 20  I = 1,ITOT                                          943
          READ (NLEG) (U(L),L=1,NQFIN), BID, (IDENT(K),K=1,NIDET)    944
          PIA  = 0.0                                                 945
          DO 10  J = 1,NACT                                          946
          JJ   = NQDEB + J                                           947
     10 PIA   = PIA + U(JJ)                                          948
          IF (PIA .LE. 1.0E-8)            PIA = 1.0 E-8              949
          WRITE (NSAV) (U(J),J=1,NQFIN), PIA, (IDENT(K),K=1,NIDET)   950
     20   CONTINUE                                                   951
          RETURN                                                     952
          END                                                        953
                                                                     954
          SUBROUTINE PUISC (ICARD, NMAX , JBASE , NQDEB , NACT , NQFIN ,NAR,  955
         1  PJ,S , BB , U , V , SOMP , TRACE , NB)                   956
C * * * * * * * * * * * * * * * * * * * * * * * * * * * * * * * * *  957
C ITERATED POWER ALGORITHM FOR DIAGONALISATION              *        958
C * * * * * * * * * * * * * * * * * * * * * * * * * * * * * * * * *  959
          DIMENSION PJ(NACT), S(NMAX,JBASE), BB(NMAX,JBASE)          960
          DIMENSION U(NQFIN) , V(NQFIN )                             961
          CALL TITLE( 0,NB)                                          962
          DO 20 J = 1,NACT                                           963
          DO 10 L = 1,JBASE                                          964
          BB(J,L) = S(J,L)                                           965
     10 S(J,L) = 0                                                   966
     20 CONTINUE                                                     967
          DO 100 IA = 1,ICARD                                        968
          READ(NB)(U(K),K=1,NQFIN) , PIA                             969
          DO 90 L = 1,JBASE                                          970
          T1 = 0.0                                                   971
          DO 30 K =1, NACT                                           972
          UUU = U(K+NQDEB)                                           973
     30 T1 = T1 + BB(K,L)*UUU /PIA                                   974
          DO 40 K = 1,NACT                                           975
          DENO = PJ(K) * SOMP                                        976
          UUU = U(K+NQDEB)                                           977
     40 S(K,L) = S(K,L) + UUU*T1/DENO                                978
          LREST = IA -(IA/NAR) * NAR                                 979
          IF(LREST*(IA - ICARD)  .NE. 0 )  GO TO 80                  980
C........ REPEATED CENTERING                                         981
          T2 = 0.0                                                   982
          DO 60 JP = 1,NACT                                          983
     60 T2 = T2 + PJ(JP)*S(JP,L)                                     984
          DO 70 J = 1,NACT                                           985
     70 S(J,L) = S(J,L) - T2                                         986
     80 CONTINUE                                                     987
     90 CONTINUE                                                     988
    100 CONTINUE                                                     989
          DOP = TRACE /( 2 * FLOAT(NACT - 1) )                       990
          DO 110 L = 1,JBASE                                         991
          DO 110 J = 1,NACT                                          992
    110 S(J,L) = S(J,L) - DOP * BB(J,L)                              993
          CALL GSMOD(NMAX,NACT,JBASE,PJ,S,KRANG,U,V)                 994
          RETURN                                                     995
          END                                                        996
                                                                     997
          SUBROUTINE CHOIX ( IX, ICARD)                              998
C * * * * * * * * * * * * * * * * * * * * * * * * * * * * * * * * *  999
C READING SELECTION CARDS FOR ROW AND COLUMN ELEMENTS        *       1000
C * * * * * * * * * * * * * * * * * * * * * * * * * * * * * * * * *  1001
          DIMENSION IX(ICARD), NBR(5)                                1002
          COMMON /ENSOR/ LEC,IMP                                     1003
          KTYP  = 5                                                  1004
          DO 10  L = 1,KTYP                                          1005
     10 NBR(L)= 0                                                    1006
C.......... READING  OF CARDS IN 80I1                                1007
          READ (LEC,1000)  (IX(I),I=1,ICARD)                         1008
          DO 20  I = 1,ICARD                                         1009
          NTYP = IX(I)                                               1010
          IF (NTYP .LE. 0)              GO TO 20                     1011
          NBR(NTYP)= NBR(NTYP) + 1                                   1012
     20   CONTINUE                                                   1013
C SUMMARY OF SELECTION OF ELEMENTS                                   1014
     70   CONTINUE                                                   1015
          WRITE (IMP,6000)                                           1016
```

217

```
      DO 80  L = 1,KTYP                                         1017
      IF (NBR(L) .LE. 0)                GO TO  80               1018
      WRITE (IMP,3000) L,   NBR(L)                              1019
 80   CONTINUE                                                  1020
      WRITE (IMP,4000)  ICARD                                   1021
      WRITE (IMP,5000) (IX(I), I=1,ICARD)                       1022
1000 FORMAT (80I1)                                              1023
3000 FORMAT (1H0,5X,4HTYPE,I3, 5X,19HNUMBER OF VARIABLES,I6)    1024
4000 FORMAT (1H0,22HINDICATOR VECTOR OF    ,I6,11H  ELEMENTS ,  1025
    1          17H IN GROUPS OF 10 /)                           1026
5000 FORMAT ( 8(1X, 10I1, 1X) )                                 1027
6000 FORMAT (1H0,21HSUMMARY OF SELECTION )                      1028
      RETURN                                                    1029
      END                                                       1030
                                                                1031
      SUBROUTINE LDPLU ( NB , NVIDI , NQEXA , IDENT , W , FMT , KFMT)  1032
C * * * * * * * * * * * * * * * * * * * * * * * * * * * * * * * * *  1033
C SUBROUTINE FOR READING NB MODIFY IF NECESSARY              *  1034
C * * * * * * * * * * * * * * * * * * * * * * * * * * * * * * * *  1035
      DIMENSION W(NQEXA) , IDENT(15), FMT(KFMT)                 1036
      IF(NVIDI .LT. 1 )                 GO TO 10                1037
      READ( NB,FMT) (IDENT(L) , L=1,NVIDI) , (W(K),K=1,NQEXA)   1038
                                        GO TO 20                1039
 10 READ (NB,FMT) (W(K),K=1,NQEXA)                              1040
 20   CONTINUE                                                  1041
      RETURN                                                    1042
      END                                                       1043
                                                                1044
      SUBROUTINE OPEN (KLE, NBAND, KOLN, LIGN, KTYP, V)         1045
C * * * * * * * * * * * * * * * * * * * * * * * * * * * * * * * * *  1046
C OPENING THE FILE ON NBAND                                  *  1047
C ATTENTION ... MINIMUM DIMENSION IN CALL  ...........V(KOLN)  *  1048
C * * * * * * * * * * * * * * * * * * * * * * * * * * * * * * * *  1049
      DIMENSION V(1), LIGN(8), LTITR(20)                        1050
      REWIND NBAND                                              1051
      READ (NBAND) (LTITR(L),L=1,20)                            1052
      READ (NBAND)   KOLN, (LIGN(K),K=1,8), KTYP                1053
      IF (KLE.EQ.2)   READ (NBAND)  (V(L),L=1,KOLN)             1054
      RETURN                                                    1055
      END                                                       1056
                                                                1057
      SUBROUTINE TITLE ( KLE, NBAND )                           1058
C * * * * * * * * * * * * * * * * * * * * * * * * * * * * * * * * *  1059
C REWINDING AND READING OF TITLE ON   NBAND                  *  1060
C PRINTING OF TITLE IF .....   KLE = 1                       *  1061
C * * * * * * * * * * * * * * * * * * * * * * * * * * * * * * * *  1062
      DIMENSION  LTITR(20)                                      1063
      COMMON  / ENSOR / LEC,IMP                                 1064
      REWIND  NBAND                                             1065
      READ (NBAND)  (LTITR(L),L=1,20), LBAND                    1066
      IF (KLE .EQ. 1) WRITE (IMP,10) NBAND, LBAND,  (LTITR(L),L=1,20)  1067
 10 FORMAT (1H0,14H INPUT FILE =    ,I5,3H  (,A4,1H),5X,20A4)   1068
      RETURN                                                    1069
      END                                                       1070
                                                                1071
      SUBROUTINE  BORNS ( N, X, XMIN, XMAX )                    1072
C * * * * * * * * * * * * * * * * * * * * * * * * * * * * * * * * *  1073
C     DETERMINE THE EXTREME VALUES  IN     VECTOR X(N)        *  1074
C          XMAX  = MAXIMUM VALUE                              *  1075
C          XMIN  = MINIMUM VALUE                              *  1076
C * * * * * * * * * * * * * * * * * * * * * * * * * * * * * * * *  1077
      DIMENSION  X(N)                                           1078
      XMIN  = X(1)                                              1079
      XMAX  = X(1)                                              1080
      IF (N .EQ. 1)         RETURN                              1081
      DO 10  I = 2,N                                            1082
      IF (X(I) .LT. XMIN)    XMIN = X(I)                        1083
      IF (X(I) .GT. XMAX)    XMAX = X(I)                        1084
 10   CONTINUE                                                  1085
      RETURN                                                    1086
      END                                                       1087
                                                                1088
```

218

```
      SUBROUTINE  EPUR4 (IDIM,ICARD, X,Y,ID, MOD, PEX, KP,KLIC,KODE,NOR, 1089
     1 NID )                                                             1090
C * * * * * * * * * * * * * * * * * * * * * * * * * * * * * * * * * * *  1091
C PROCEDURE TO BRING BACK THE POINTS MORE THAN PEX STANDARD DEVIATIONS* 1092
C TO THE PERIPHERY OF THE GRAPH.ON THE OUTPUT THE KP POINTS WHICH    *  1093
C HAVE BEEN MODIFIED ARE IDENTIFIED BY THE NUMBERS IN KLIC( )        *  1094
C KODE=1 IF ICARD + 1 IS GREATER THAN IDIM OR THERE ARE MORE THAN    *  1095
C 264 POINTS ON THE GRAPH.(THIS IS THE MAXIMIUM DIMENSION OF KLIC)   *  1096
C IF MOD = 1 THE IDENTIFIERS ARE IN A1. IF MOD = 4 THE IDENTIFIERS   *  1097
C ARE IN A4.                                                         *  1098
C NOTE X(*) AND Y(*) ARE DESTROYED IF KP.NE.0                        *  1099
C * * * * * * * * * * * * * * * * * * * * * * * * * * * * * * * * * * *  1100
      DIMENSION  X(IDIM),Y(IDIM),ID(IDIM,NID),IDA(3)                     1101
      DIMENSION  KLIC(264)                                               1102
      COMMON /ENSOR /LEC,IMP                                             1103
      DATA SEUIL / 1.0 E-07 /                                           1104
      DATA LBLAN/4H    /                                                 1105
      IF (PEX .LE. 0.0)             RETURN                               1106
      KODE  = 0                                                          1107
      N1    = ICARD + 1                                                  1108
      IF (N1 .GT. IDIM)             GO TO  100                           1109
      X(N1) = 0.0                                                        1110
      Y(N1) = 0.0                                                        1111
      IF( NOR .NE. 1 )              N1 = ICARD                           1112
      SX    = 0.0                                                        1113
      SY    = 0.0                                                        1114
      DO 10  I = 1,ICARD                                                 1115
      SX    = SX + X(I)*X(I)                                             1116
   10 SY    = SY + Y(I)*Y(I)                                             1117
      SX    = SQRT(SX/FLOAT(ICARD))                                      1118
      SY    = SQRT(SY/FLOAT(ICARD))                                      1119
      PX    = PEX*SX                                                     1120
      PY    = PEX*SY                                                     1121
      KP    = 0                                                          1122
      PIN   = PEX + 20.0                                                 1123
      DO 30  I = 1,ICARD                                                 1124
      IF (ABS(X(I)) .GT. PX)        GO TO  20                            1125
      IF (ABS(Y(I)) .LE. PY)        GO TO  30                            1126
   20 KP    = KP + 1                                                     1127
      IF (MOD.NE.1 .AND. KP.EQ.1)   WRITE(IMP,120)  PEX                  1128
      IF (MOD.NE.1 .AND. KP.EQ.1)   WRITE(IMP,130)                       1129
      IF (KP .GT. 264)              GO TO  100                           1130
      KLIC(KP)= I                                                        1131
      DO 25 K = 1,3                                                      1132
      IDA(K) = LBLAN                                                     1133
      IF( NID .LT. K )    GO TO 25                                       1134
      IDA(K) = ID(I,K)                                                   1135
   25 CONTINUE                                                           1136
      IF (MOD .NE. 1)          WRITE(IMP,110) (IDA(L),L=1,3),X(I),Y(I)   1137
      X(I)  = X(I) / PIN                                                 1138
      Y(I)  = Y(I) / PIN                                                 1139
   30   CONTINUE                                                         1140
      IF (KP .EQ. 0)                GO TO  90                            1141
      IF (MOD .EQ. 4)               WRITE(IMP,130)                       1142
      CALL BORNS (N1,Y,YMIN,YMAX)                                        1143
      CALL BORNS (N1,X,XMIN,XMAX)                                        1144
      IF (XMIN .EQ. 0.)             XMIN = SEUIL                         1145
      IF (XMAX .EQ. 0.)             XMAX = SEUIL                         1146
      T1    = YMAX / XMAX                                                1147
      T2    = YMAX / XMIN                                                1148
      T3    = YMIN / XMIN                                                1149
      T4    = YMIN / XMAX                                                1150
      DO 80  K = 1,KP                                                    1151
      KLK = KLIC(K)                                                      1152
      U = X(KLK)                                                         1153
      V = Y(KLK)                                                         1154
      IF (U .EQ. 0.)                U = SEUIL                            1155
      T     = V / U                                                      1156
      A     = YMAX                                                       1157
      IF(U.GT.0.0 .AND.(T4.LT.T .AND. T.LT.T1))  GO TO  40              1158
      IF(U.LT.0.0 .AND.(T2.LT.T .AND. T.LT.T3))  GO TO  50              1159
      IF (V .LT. 0.)                A = YMIN                             1160
      V     = A                                                          1161
      IF (T .EQ. 0.)                T = SEUIL                            1162
      U     = A / T                                                      1163
                                    GO TO  70                           1164
```

219

```
   40 A       = XMAX                                                      1165
                                   GO TO  60                              1166
   50 A       = XMIN                                                      1167
   60 U       = A                                                         1168
      V       = A * T                                                     1169
   70 CONTINUE                                                            1170
         IF( V .GT. YMAX)              V = YMAX                           1171
         IF( U .GT. XMAX)              U = XMAX                           1172
      X(KLK) = U                                                          1173
      Y(KLK) = V                                                          1174
   80   CONTINUE                                                          1175
         IF (KP.NE.0 .AND. MOD.EQ.1)   WRITE(IMP,140) KP , PEX            1176
   90   CONTINUE                                                          1177
      RETURN                                                              1178
  100 KODE  = 1                                                           1179
  110 FORMAT (1H ,2H *,3A4,1X,1H*,2(F13.5,8X,1H*) )                       1180
  120 FORMAT (///1H  ,      25X,9HATTENTION,//1H ,                        1181
     1    39HTHE POINTS BELOW WERE MORE THAN       ,F5.1,                 1182
     2    29H STD.DEVIATIONS FROM CENTER /1H ,14HTHEY HAVE BEEN,          1183
     3    35H BROUGHT BACK TO PERIPHERY OF GRAPH      // )                1184
  130 FORMAT (1H ,30(2H * ) )                                             1185
  140 FORMAT (1H ,5X, I4, 18H  POINTS MORE THAN,F5.1,14H STD.DEVS.  ,     1186
     1  52H FROM THE ORIGIN HAVE BEEN BROUGHT BACK TO PERIPHERY)          1187
      RETURN                                                              1188
      END                                                                 1189
                                                                          1190
      SUBROUTINE HPLAN (IDIM,ICARD,JX,JY,X,Y,ID,MOD,NLIGN,NPAGE,PEX,NOR,  1191
     1 NID ,NBAND)                                                        1192
C * * * * * * * * * * * * * * * * * * * * * * * * * * * * * * * * * * *   1193
C   DISPLAY   OF ICARD POINTS, ON  NLIGN- ROWS, AND NPAGE-PAGES       *   1194
C (IF NLIGN=0 NLIGN SET AUTOMATICALLY. NPAGE =1 TO 8)                 *   1195
C NOTE IF MODE=1 THERE CANNOT BE MORE THAN 3 PAGES UNLESS THE ARRAY   *   1196
C IS REDIMENSIONED                                                    *   1197
C COORDINATES X(*) ON  AXIS JX HORIZONTAL, AND Y(*) ON JY VERTICAL    *   1198
C      IDENTIFIERS   IN ID(*,*), IN A1 IF MOD=1,  IN A4 IF MOD=4      *   1199
C POINTS MORE THAN PEX STANDARD DEVIATIONS ARE BROUGHT BACK TO        *   1200
C THE PERIPHERY                                                       *   1201
C IF NOR=1 THE ORIGIN OF THE AXIS IS PLACED ON THE GRAPH              *   1202
C   NID=1,2,3   LENGTH IDENTIFIER IN NUMBER OF A4                     *   1203
C      NOTE      / X(*), Y(*), ID(ICARD+1) ARE DESTROYED /            *   1204
C   DISPLAY   SUPPRESSED IF MORE OF 264 POINTS ON     GRAPH,          *   1205
C ONLY THE FIRST 100 DUPLICATED ARE REFERENCED BY THEIR CO-ORDINATES  *   1206
C * * * * * * * * * * * * * * * * * * * * * * * * * * * * * * * * * * *   1207
      DIMENSION  X(IDIM) , Y(IDIM) , ID(IDIM,NID) , LD1(100) , LD2(100)   1208
      DIMENSION  EX(48),KLIC(371),KLAC(371),LA(6),MA(6),NA(6)             1209
      COMMON /ENSOR /LEC,IMP                                              1210
      DATA MA/1H-,1H ,1HI ,1HI,1H.,1H+/,LDA2/4H    /,LDB2/4H /            1211
      DATA NA/4H----,4H    ,4HI   ,4H I,4H . ,4H + /                      1212
      IF (NPAGE .EQ. 0)             NPAGE = 1                             1213
      N1    = ICARD                                                       1214
      REWIND NBAND                                                        1215
      IF (NOR .EQ. 1)               N1 = ICARD + 1                        1216
      IF (N1 .GT. IDIM)             WRITE (IMP,1130)                      1217
      WRITE (IMP,1110)  ICARD, JX, JY, JX, JY                            1218
      CALL EPUR4 ( IDIM,ICARD, X,Y,ID, MOD, PEX, KP, KLIC,KODE,NOR ,NID)  1219
      IF (KODE .EQ. 1)             GO TO 160                              1220
      WRITE (IMP,1000)                                                    1221
      DO 10  K = 1,6                                                      1222
      LA(K) = NA(K)                                                       1223
   10   IF  (MOD .EQ. 1)           LA(K) = MA(K)                          1224
      IF (NOR .NE. 1)             GO TO 20                                1225
      ID(N1,1)= LA(6)                                                     1226
      IF(NID .GT. 1) ID(N1,2) = LA(2)                                     1227
      IF(NID .GT. 2) ID(N1,3) =LA(2)                                      1228
      X(N1) = 0.0                                                         1229
      Y(N1) = 0.0                                                         1230
   20 CALL BORNS (N1,X,XMIN,XMAX)                                         1231
      CALL BORNS (N1,Y,YMIN,YMAX)                                         1232
      IF (MOD .EQ. 1)            K1 = 123                                 1233
      IF (MOD .EQ. 4)            K1 =  30                                 1234
      KFIN  = K1 * NPAGE                                                  1235
      NL    = NLIGN                                                       1236
      FC    = KFIN                                                        1237
      FPAGE = NPAGE                                                       1238
```

                                    220

```
      IF (NL .NE. 0)                       GO TO  30              1239
      NL   = ((YMAX-YMIN) / (XMAX-XMIN))*FPAGE*74.0               1240
 30   IF (NL .LE. 12)                   NL = 12                   1241
      FL   = NL-1                                                 1242
      S    = (XMAX-XMIN) / FC                                     1243
      T    = (YMAX-YMIN) / FL                                     1244
      NINT = 5*NPAGE + 1                                          1245
      ESPX = FC / (5.0*FPAGE)                                     1246
      DO 40  J = 1,NINT                                           1247
 40   EX(J) = XMIN + S*ESPX*FLOAT(J-1)                            1248
      KKO  = 0.50001 - (XMIN/S)                                   1249
      LLO  = 1.00000 + ABS(YMAX/T)                                1250
      DO 50  I = 1,N1                                             1251
      K    = (X(I) - XMIN)/S + 0.500001                           1252
      L    = (YMAX - Y(I))/T + 1.00000                            1253
      IF (K .EQ. 0)                     K = 1                     1254
      IF (L .EQ. 0)                     L = 1                     1255
      X(I) = K + 0.0001                                           1256
 50   Y(I) = L + 0.0001                                           1257
      J    = 0                                                    1258
      DO 110 LL = 1,NL                                            1259
      EY   = YMAX - T*FLOAT(LL-1)                                 1260
      DO 60  KK = 1,KFIN                                          1261
      KLIC(KK)= 0                                                 1262
      KLAC(KK)= LA(2)                                             1263
      IF (LL.EQ.1 .OR. LL.EQ.NL)    KLAC(KK) = LA(1)              1264
      IF (KK.EQ.KKO .OR. LL.EQ.LLO) KLAC(KK) = LA(5)              1265
 60   CONTINUE                                                    1266
      KLAC(1) = LA(3)                                             1267
      KLAC(KFIN)= LA(4)                                           1268
      KLAC(KFIN + 1) = LA(2)                                      1269
      KLIC(KFIN + 1) = 0                                          1270
      KLIC(KFIN+2)=0                                              1271
      DO 90  I = 1,N1                                             1272
      L     = Y(I)                                                1273
      IF (L .NE. LL)                    GO TO  90                 1274
      K     = X(I)                                                1275
      IF (KLIC(K))                      80, 70, 80                1276
 70   KLIC(K)= I                                                  1277
                                        GO TO  90                 1278
 80   IK    = KLIC(K)                                             1279
      J     = J + 1                                               1280
      IF (MOD .EQ. 1)                   GO TO  90                 1281
      IF (J .GT. 100)                   GO TO  90                 1282
      LD1(J)= IK                                                  1283
      LD2(J)= I                                                   1284
 90   CONTINUE                                                    1285
      DO 100  KK = 1,KFIN                                         1286
      IK    = KLIC(KK)                                            1287
      IF(IK .EQ.0 )   GO TO 100                                   1288
                      KLAC(KK) = ID(IK,1)                         1289
      IF(KLIC(KK+1) .EQ.0 .AND .NID. GT. 1 ) KLAC(KK+1)=ID(IK,2)  1290
      IF( (KK+1)    .EQ. KFIN)  GO TO 100                         1291
      IF(KLIC(KK+2) .EQ.0 .AND .NID. GT. 2 ) KLAC(KK+2)=ID(IK,3)  1292
 100  CONTINUE                                                    1293
      KFI1 = KFIN + 1                                             1294
      WRITE (NBAND)  EY, (KLAC(K), K=1,KFI1)                      1295
 110  CONTINUE                                                    1296
      KB   = 0                                                    1297
      KBB  = 1                                                    1298
      DO 135  KPT = 1,NPAGE                                       1299
      REWIND NBAND                                                1300
      KA    = KB + 1                                              1301
      KB    = KA + K1 - 1                                         1302
      IF( KB .EQ. KFIN .AND. KPT.GT. 1 .AND. MOD.NE.1)  KB = KFIN + 1   1303
      KAA   = KBB + 1                                             1304
      KBB   = KAA + 4                                             1305
      DO 130 LL = 1,NL                                            1306
      READ (NBAND) EY, (KLAC(K), K=1,KFI1)                        1307
      IF (MOD .NE. 1)                   GO TO 120                 1308
      IF (KPT .EQ. 1) WRITE(IMP,1010) EY, (KLAC(K),K=KA,KB)       1309
      IF (KPT .NE. 1) WRITE(IMP,1030) (KLAC(K),K=KA,KB)           1310
                                        GO TO 130                 1311
 120     IF (KPT .EQ. 1) WRITE (IMP,1020) EY,(KLAC(K),K=KA,KB)    1312
```

221

```
         IF (KPT .NE. 1) WRITE (IMP,1040) (KLAC(K),K=KA,KB)         1313
130  CONTINUE                                                        1314
         IF (KPT.EQ.1)   WRITE (IMP,1050) (EX(K),K=1,6)              1315
         IF (KPT .GT. 1) WRITE (IMP,1060) (EX(K),K= KAA , KBB )      1316
         IF (KPT .LT. NPAGE)   WRITE (IMP,1000)                      1317
135  CONTINUE                                                        1318
         IF (J.EQ.0 .OR. MOD.EQ.1)        GO TO  150                 1319
     WRITE (IMP,1070)                                                1320
     WRITE (IMP,1080)                                                1321
     LTOT = MINO(J,100)                                              1322
         DO 140 L=1,LTOT                                             1323
     IK    = LD1(L)                                                  1324
     I     = LD2(L)                                                  1325
     LDA   = ID(IK,1)                                                1326
     LDB   = ID(I,1)                                                 1327
     IF(NID .GT. 1 ) LDA2 = ID(IK,2)                                 1328
     IF(NID .GT. 1 ) LDB2 = ID(I,2)                                  1329
     XD1   = X(IK)*S + XMIN                                          1330
     YD1   = YMAX - Y(IK)*T                                          1331
     XD2   = X(I)*S + XMIN                                           1332
     YD2   = YMAX - Y(I)*T                                           1333
140  WRITE (IMP,1090) LDA,LDA2,XD1,YD1,LDB,LDB2,XD2,YD2             1334
     WRITE (IMP,1080)                                                1335
150  CONTINUE                                                        1336
         IF (J .NE. 0)                   WRITE(IMP,1100)   J         1337
         RETURN                                                      1338
160  WRITE (IMP,1120)                                                1339
1000 FORMAT (1H1)                                                    1340
1010 FORMAT (1H ,F8.3,1X,123A1)                                      1341
1020 FORMAT (1H ,F8.3,2X,30A4)                                       1342
1030 FORMAT (1H ,123A1)                                              1343
1040 FORMAT (1H ,31A4)                                               1344
1050 FORMAT (1H ,2X,5(F10.3,14X),F10.3)                              1345
1060 FORMAT (1H ,18X,4(F10.3,14X),F10.3)                             1346
1070 FORMAT (1H0/,1H ,50X,35HMULTIPLE POINTS  ( 100   MAXIMUM ) ,/,  1347
    1        /1H ,2(8H POINT *,8X,8HABSCISSA,7X,1H*,8X,8HORDINATE,7X,1348
    2 1H*,10X)/1H ,8H SEEN  *,2(8X,9HAPPROX.  ,6X,1H*),10X,          1349
    3          8HHIDDEN *,2(8X,9HAPPROX.  ,6X,1H*) )                 1350
1080 FORMAT (1H ,2(28(2H *),10X) )                                   1351
1090 FORMAT (1H ,2(2H * ,2A4,2H *,2(F13.2,8X,1H*),10X)  )            1352
1100 FORMAT ( / 1H ,30H NUMBER OF DOUBLE POINTS =     ,I5, /)        1353
1110 FORMAT (///1H ,25X,22HGRAPH OF             ,I6, 8H POINTS ,     1354
    1  14H ON THE  AXES ,I2,6H AND  , I2,/1H ,130(1H-)//,1H ,30X,    1355
    2  4HAXIS,I2,14H /HORIZONTAL ,10X, 4HAXIS,I2,12H /VERTICAL // )  1356
1120 FORMAT (///1H ,10X,17HDISPLAY  REJECTS /                        1357
    1  1H ,10X,36H  MORE THAN 264 POINTS ON THE PAGE /1H1 )          1358
1130 FORMAT (/1H ,47H(NOTE     DIMENSION PROBLEM IN S/R HPLAN)   /)  1359
     RETURN                                                          1360
     END                                                             1361
                                                                     1362
     FUNCTION  SEN3A ( BIDON )                                       1363
C * * * * * * * * * * * * * * * * * * * * * * * * * * * * * * *      1364
C UNIFORM PSEUDO-RANDOM NUMBER GENERATOR (0,1)              *        1365
C  REFERENCE /K.D.SENNE/J. STOCHASTICS/ VOL 1,NO 3 (1974),PP.215-38/ * 1366
C * * * * * * * * * * * * * * * * * * * * * * * * * * * * * * *      1367
C.......... CONSTANTS                                                1368
     DATA M12/ 4096 /                                                1369
     DATA F1/2.44140625E-04/,F2/5.96046448E-08/,F3/1.45519152E-11/   1370
     DATA J1/ 3823 /       ,J2/ 4006 /       ,J3/ 2903 /             1371
     DATA I1/ 3823 /       ,I2/ 4006 /       ,I3/ 2903 /             1372
C..........CALCULATION OF CONGRUENCE WITH SEGMENTATION  OF NUMBERS   1373
     K3 = I3*J3                                                      1374
     L3 = K3/M12                                                     1375
     K2 = I2*J3 + I3*J2 + L3                                         1376
     L2 = K2/M12                                                     1377
     K1 = I1*J3 + I2*J2 + I3*J1 + L2                                 1378
     L1 = K1/M12                                                     1379
     I1 = K1 - L1*M12                                                1380
     I2 = K2 - L2*M12                                                1381
     I3 = K3 - L3*M12                                                1382
     SEN3A = F1*FLOAT(I1)+ F2*FLOAT(I2)+F3*FLOAT(I3)                 1383
     RETURN                                                          1384
     END                                                             1385
                                                                     1386
```

222

References

Anderberg, M. R. (1973). *Cluster Analysis for Applications*, Academic Press, New York.

Anderson, T. W. (1958). *An Introduction to Multivariate Statistical Analysis*. Wiley, New York.

Anderson, T. W. (1963). Asymptotic Theory for Principal Components Analysis. *Ann. Math. Stat.*, Vol. 34, pp. 122–148.

Ball, G. H., and Hall, D. J. (1965). Isodata, A Novel Method of Data Analysis and Pattern Classification. Technical report 5RI. Project 5533, Stanford Research Institute, Menlo Park, California.

Ball, G. H., and Hall, D. J. (1967). A Clustering Technique for Summarizing Multivariate Data. *Behav. Sci.*. No. 12, pp. 153–155.

Barnett, V. (1981). *Interpreting Multivariate Data*. Wiley, New York.

Bartlett, M. S. (1951). The effect of standardization on χ^2 approximation in factor analysis (with an appendix by W. Lederman). *Biometrika*, Vol. 38, pp. 337–344.

Beale, E. M. L. (1969). Euclidean Cluster Analysis. *Bull. I.S.I.*, Vol. 43, Book 2, pp. 92–94.

Benzecri, J. P. (1964). Cours de Linguistique Mathématique. Publication Mimeo, Faculté des Sciences, Rennes, France.

Benzecri, J. P. (1969a). Statistical Analysis as a Test to Make Patterns Emerge from Clouds. In *Methodologies of Pattern Recognition* (S. Watanabe, Ed.). Academic Press, New York, pp. 35–74.

Benzecri, J. P. (1969b). Approximation Stochastique dans une Algèbre Normée Non Commutative. *Bull. Soc. Math. Fr.*, No. 97, pp. 225–241.

Benzecri, J. P. (1972). Sur l'Analyse des Tableaux Binaires Associés à une Correspondance Multiple. Note Mimeo, Lab. Stat. Math., Université Pierre et Marie Curie, Paris.

Benzecri, J. P. (1973). *L'Analyse des Données, Tome 1: La Taxinomie, Tome 2: L'Analyse des Correspondances*. Dunod, Paris (2nd ed. 1976).

Benzecri, J. P. (1974). La Place de l'a priori. In *Encyclopaedia Universalis* (Section: Organum). Paris.

Benzecri, J. P. (1976). Histoire et Préhistoire de l'Analyse des Données. *Cah. Anal. Données*, Vol. 1, No. 1–4 (published in book form: 1982, Dunod, Paris).

Benzecri, J. P. (1979). Sur le Calcul des Taux d'Inertie dans l'Analyse d'un Questionnaire. *Cah. Anal. Données*, Vol. 4, No. 3, pp. 378–379.

Benzecri, J. P. (1982). Construction d'une Classification Ascendante Hiérarchique par la Recherche en Chaîne des Voisins Réciproques. *Cah. Anal. Données*, Vol. 7, No. 2, pp. 209–218.

Bouroche, J. M., and Saporta G. (1980). *L'Analyse des Données*. Presses Universitaires de France, Paris.

Brillouin, L. (1959). *La Science et la Théorie de l'Information*. Masson, Paris.

Burt, C. (1950). The Factorial Analysis of Qualitative Data. *J. Stat. Psychol.*, Vol. 3, No. 3, pp. 166–185.

Caillez, F., and Pages, J. P. (1976). *Introduction à l'Analyse des Données*. S.M.A.S.H., Paris.

Carroll, J. D. (1968). Generalization of Canonical Correlation to Three or More Sets of Variables. *Proc. Amer. Psychol. Assoc.*, pp. 227–228.

Cattell, R. B. (1978). *The Scientific Use of Factor Analysis in Behavioral and Life Science.* Plenum Press, New York.

Cazes, P. (1976). Etude des Propriétés Extrêmales des Facteurs Issus d'un Sous-Tableau d'un Tableau de Burt. Note Mimeo, Lab. Stat. Math., Université Pierre et Marie Curie, Paris.

Choudary Hanumara, R., and Thompson, W. A. (1968). Percentage Point of the Extreme Roots of the Wishart Matrix. *Biometrika*, No. 55, pp. 505–512.

Clemm, D. S., Krishnaiah, P. R., and Waikar, V. B. (1973). Tables of the Extreme Roots of a Wishart Matrix. *J. Statist. Comput. Simul.*, Vol. 2, pp. 65–92.

Clint, M., and Jennings, A. (1970). The Evaluation of Eigenvalues and Eigenvectors of Real Symmetric Matrices by Simultaneous Iteration. *Comput. J.*, Vol. 13, pp. 76–80.

Cormack, R. M. (1971). A Review of Classification. *J. Roy. Statist. Soc. A*, Vol. 134, Part 3.

Corsten, L. C. A. (1976). Matrix Approximation, A Key to Application of Multivariate Methods. In *Proc. 9th Int. Biometric Conf.*, Vol. 1, pp. 61–77. Raleigh, North Carolina.

Davis, A. W. (1972). On the ratios of the individual latent roots to the trace of a Wishart matrix. *J. Mult. Anal.*, Vol. 2, pp. 440–443.

Davis, A. W. (1977). Asymptotic Theory for Principal Components Analysis: Nonnormal Case. *Aust. J. Statist.*, Vol. 19, pp. 206–212.

Dempster, A. P. (1969). *Elements of Continuous Multivariate Analysis*. Addison-Wesley, Reading.

Diday, E. (1971). La Méthode des Nuées Dynamiques. *Rev. Statist. Appl.*, Vol. 19, No. 2, pp. 19–34.

Diday, E. (1972). Optimisation en Classification Automatique et Reconnaissance des Formes. *Rev. Fr. Inf. Rech. Opér.*, 6th year (November 1972), pp. 61–95.

Diday, E. (1974). Classification Automatique Séquentielle pour Grands Tableaux. *Rev. Fr. Inf. Rech. Opér.*, 9th year (March 1975), pp. 1–29.

Diday, E., et al. (1980). *Optimisation en Classification Automatique* (Tome 1 et 2). INRIA, Rocquencourt.

Eckart, C., and Young, G. (1936). The Approximation of One Matrix by Another of Lower Rank. *Psychometrika*, No. 1, pp. 211–218.

Eckart, C., and Young, G. (1939). A Principal Axis Transformation for Non-Hermitian Matrices. *Bull. Amer. Math. Assoc.*, Vol. 45, pp. 118–121.

Efron, B. (1979). Bootstrap Methods: Another Look at the Jackknife. *Ann. Statist.*, Vol. 7, pp. 1–26.

Escofier-Cordier, B. (1965). L'Analyse des Correspondances. Thesis published in 1969 in *Cah. Bur. Univ. Rech. Opér.*, No. 13.

Escofier, B., and Leroux, B. (1972). Etude de Trois Problèmes de Stabilité en Analyse Factorielle. *Publ. Inst. Statist. Univ. Paris*, Vol. 11, pp. 1–48.

Escoufier, Y. (1970). Echantillonnage dans une Population de Variables Alétoires Réelles. *Publ. Inst. Statist. Univ. Paris*, Vol. 19, Fas. 4, pp. 1–47.

Fenelon, J. P. (1981). *Qu'est ce que l'Analyse des Données*. Lefonen, Paris.

Fisher, R. A. (1936). The Use of Multiple Measurements in Taxonomic Problems. *Ann. Eugen.*, Vol. 7, pp. 179–188.

Fisher, R. A. (1939). The Sampling Distribution of Some Statistics Obtained from Non Linear Equations. *Ann. Eugen.*, Vol. 9, pp. 238–249.

Fisher, R. A. (1940). The Precision of Discriminant Functions. *Ann. Eugen.*, Vol. 10, pp. 422–429.

Fisher, R. A., and Yates F. (1949), *Statistical Tables for Biological, Agricultural and Medical Research*. Hafner Publishing Company, New York.

Fisher, W. D. (1958). On Grouping for Maximum Homogeneity. *J. Amer. Statist. Assoc.*, No. 53, pp. 789–798.

Florek, K. et al. (1951). Sur la Liaison et la Division des Points d'un Ensemble Fini. *Colloq. Math.*, Vol. 2, pp. 282–285.

Forgy, E. W. (1965). Cluster Analysis of Multivariate Data: Efficiency Versus Interpretability of Classifications. *Biometrics*, Vol. 21, pp. 768–769.

Friedman, H. P., and Rubin, J. (1967). On Some Invariant Criteria for Grouping Data. *J. Amer. Statist. Assoc.*, Vol. 62, pp. 1159–1178.

Gabriel, K. R. (1981). Biplot Display of Multivariate Matrices for Inspection of Data and Diagnosis. In *Interpreting Multivariate Data* (V. Barnett, (Ed.), Wiley, Chichester, pp. 147–173.

Gifi, A. (1981). *Non-Linear Multivariate Analysis*. Leiden, Dept. of Datatheory.

Girshick, M.A. (1939). On the Sampling Theory of Roots of Determinantal Equations. *Ann. Math. Statist.*, Vol. 10, pp. 203–224.

Gordon, A. D. (1981). *Classification*. Chapman and Hall, London.

Gourlay, A. R., and Watson, G. A. (1973). Computational Methods for Matrix Eigen Problems. Wiley, New York.

Gower, J. C. (1966). Some Distance Properties of Latent and Vector Methods Used in Multivariate Analysis. *Biometrika*, Vol. 53, pp. 325–328.

Gower, J. C. and Ross G. (1969). Minimum Spanning Trees and Single Linkage Cluster Analysis. *Appl. Statist.*, Vol. 18, pp. 54–64.

Greenacre, M. J. (1981). Practical Correspondence Analysis. in *Interpreting Multivariate Data* (V. Barnett, Ed.), Wiley, Chichester, pp. 119–146.

Harman, H. H. (1976). *Modern Factor Analysis* (3rd ed.), Chicago University Press.

Hartigan, J. A. (1975). *Clustering Algorithms*. Wiley, New York.

Hayashi, C. (1950). On the Quantification of Qualitative Data from the Mathematico-Statistical Point of View. *Ann. Inst. Statist. Math.*, Vol. 2, Tokyo.

Hill, M. O. (1973). Reciprocal Averaging: An Eigenvector Method of Ordination. *J. Ecol.*, Vol. 61, pp. 237–251.

Hill, M. O. (1974). Correspondence Analysis: A Neglected Multivariate Method. *Appl. Statist.*, No. 3, pp. 340–354.

Horst, P. (1961). Relation Among *m* Sets of Measures. *Psychometrika*, Vol. 26, pp. 129–149.

Horst, P. (1965). *Factor Analysis of Data Matrices*. Holt, Rinehart, Winston, New York.

Hotelling, H. (1933). Analysis of a Complex of Statistical Variables into Principal Components. *J. Educ. Psychol.*, Vol. 24, pp. 417–441, and pp. 498–520.

Hotelling, H. (1936). Relation Between Two Sets of Variables. *Biometrika*, Vol. 28, pp. 129–149.

Householder, A. S. (1953). *Principles of Numerical Analysis*. McGraw-Hill, New York.

Hsu, P. L. (1939). On the Distribution of the Roots of Certain Determinantal Equations. *Ann. Eugen.*, Vol. 9, pp. 250–258.

Jambu, M., and Lebeaux, M.O., (1983). *Cluster Analysis and Data Analysis*. North-Holland, Amsterdam.

Jeffreys, H. (1946). An Invariant Form for the Prior Probability in Estimation Problems. *Proc. Roy. Soc. (A)*, Vol. 186, pp. 453–461.

Johnson, S. C. (1967). Hierarchical Clustering Schemes. *Psychometrika*, Vol. 32, pp. 241–254.

Jousselin, B. (1972). Les Choix de Consommation et les Budgets des Ménages. *Consommation*,

No. 1, pp. 41–72.

Juan, J. (1982). Programme de Classification Hiérarchique par l'Algorithme de la Recherche en Chaîne des Voisins Réciproques. *Cah. Anal. Données*, Vol. 7, No. 2, pp. 219–226.

Kendall, M. G., and Stuart, A. (1961). *The Advanced Theory of Statistics*, Vol. 2. Griffin, London.

Kettenring, J. R. (1971). Canonical Analysis of Several Sets of Variables. *Biometrika*, Vol. 58, No. 3, pp. 433–450.

Krishnaiah, P. R., and Chang, T. C. (1971). On the Exact Distribution of the Extreme Roots of the Wishart and Manova Matrices. *J. Mult. Anal.*, Vol. 1, No. 1, pp. 108–116.

Krishnaiah, P. R., and Waikar, V. B. (1971). Exact Joint Distribution of Any Few Ordered Roots of a Class of Random Matrices. *J. Mult. Anal.*, Vol. 1, No. 3, pp. 308–315.

Kruskal, J. B. (1956). On the Shortest Spanning Subtree of a Graph and the Travelling Salesman Problem. *Proc. Amer. Math. Soc.*, Vol. 7, pp. 48–50.

Kshirsagar, A. M. (1972). *Multivariate Analysis*. Marcel Dekker, Inc., New York.

Kullback, S. (1959). *Information Theory and Statistics*. Wiley, New York.

Lancaster, H. O. (1963). Canonical Correlation and Partition of X^2. *Quart. J. Math.*, Vol. 14, pp. 220–224.

Lancaster, H. O. (1969). *The Chi-Squared Distribution*. Wiley, New York.

Lance, G. N., and Williams, W. T. (1967). A General Theory of Classification Sorting Strategies. II Clustering systems. *Comput. J.*, Vol. 10, pp. 271–277.

Lawley, N. N., and Maxwell, A. E. (1970). *Factor Analysis as a Statistical Method*. Butterworths, London.

Lebart, L. (1969). Introduction à l'Analyse des Données. *Consommation*. No. 4, pp. 65–86.

Lebart, L. (1974). On the Benzecri's Method for Finding Eigenvector by Stochastic Approximation (The Case of Binary Data). In *Proc. Comput. Statist.* pp. 202–211. Physica Verlag, Vienna.

Lebart, L. (1975a). L'Orientation du Dépouillement de Certaines Enquêtes par l'Analyse des Correspondances Multiples. *Consommation*, No. 2, pp. 73–96.

Lebart, L. (1975b). Validité des Résultats en Analyse des Données. Report CREDOC-DGRST, Paris.

Lebart, L. (1976). The Significancy of Eigenvalues Issued from Correspondence Analysis. In *Proc. Comput. Statist.*, pp. 38–45. Physica Verlag, Vienna.

Lebart, L. (1982). Exploratory Analysis of Large Sparse Matrices with Application to Textual Data. In *COMPSTAT 1982*, pp. 67–76. Physica Verlag, Vienna.

Lebart, L. (1983). Exploratory Analysis of Some Large Sets of Data: A Critical Balance. In *Proc. 44th Session of the I.S.I.*, Madrid.

Lebart, L., and Fenelon, J. P. (1971). *Statistique et Informatique Appliquées*. Dunod, Paris (3rd ed., 1975).

Lebart, L., and Houzel van Effentere, Y. (1980). Le Système d'Enquête sur les Conditions de Vie et Aspirations des Français. *Consommation*, No. 1, 1980.

Lebart, L., and Morineau, A. (1982). *SPAD, Système Portable pour L'Analyse des Données*. CESIA, Paris.

Lebart, L., Morineau, A., and Fenelon, J. P. (1979). *Traitement des Données Statistiques, Méthodes et Programmes*. Dunod, Paris.

Lebart, L, and Tabard, N. (1973). Recherches sur la Description Automatique des Données Socio-Economiques. Report CORDES-CREDOC, Paris.

Leclerc, A. (1975). L'Analyse des Correspondances sur Juxtaposition de Tableaux de Contingence. *Rev. Statist. Appl.*, Vol. 23, No. 3, pp. 5–16.

McKeon, J. J. (1966). Canonical Analysis: Some Relations Between Canonical Correlation Factor Analysis, Discriminant Analysis, and Scaling Theory. Psychometric Monograph,

Vol. 13.

MacQueen, J. (1967). Some Methods for Classification and Analysis of Multivariate Observations. In *Proc. 5th Berkeley Symp. 1965*, pp. 281–297.

Mahalanobis, P. C. (1936). On the Generalized Distance in Statistics. *Proc. Nat. Inst. Sci.—India*, Vol. 12, pp. 49–55.

Masson, M. (1974). Analyse Non Linéaire de Données. *C.R. Acad. Sci.*, Vol. 278, (March 11).

MacQuitty, L. L. (1977). Single and Multiple Hierarchical Classification by Reciprocal Pairs and Rank Order Types. *Educ. Psychol. Meas.*, Vol. 26, pp. 253–265.

Mehta, M. L. (1960). On the Statistical Properties of the Level Spacing in Nuclear Spectra. *Nucl. Phys.*, Vol. 18, pp. 395–419.

Mehta, M. L. (1967). *Random Matrices and the Statistical Theory of Energy Levels*. Academic Press, New York.

Miller, R. G. (1974). The Jackknife: A Review. *Biometrika*, Vol. 61, pp. 1–15.

Mood, A. M. (1951). On the Distribution of the Characteristic Roots of Normal Second Moment Matrices. *Ann. Math. Statist.*, Vol. 22, pp. 266–273.

Morineau, A. (1982). Choice of Methods and Algorithms for Statistical Treatments of Large Arrays of Data. In: COMPSTAT 1982, pp. 342–347. Physica Verlag, Vienna.

Morineau, A. (1973). Aspects Descriptifs de Certaines Méthodes d'Analyse des Données. Report mimeo, CEPREMAP, Paris.

Morineau, A. (1983). Etudes de Stabilité en Analyse en Composantes Principales. Technical Bulletin. Vol. 1, pp. 9–12. CESIA, Paris.

Morrison, D. F. (1976). *Multivariate Statistical Methods*. McGraw-Hill, New York.

Muirhead, R. J. (1982). *Aspects of Multivariate Statistical Theory*. Wiley, New York.

Mulaik, S. A. (1972). *The Foundations of Factor Analysis*. McGraw-Hill, New York.

Nakache, J. P. (1973). Influence du Codage des Données en Analyse Factorielle des Correspondances, Etude d'un Exemple Pratique Médical. *Rev. Statist. Appl.*, Vol. 21, No. 2.

Nakhle, F. (1976). Sur l'Analyse d'un Tableau de Notes Dédoublées. *Cah. Anal. Données*, Vol. 1, No. 3, pp. 243–257.

Nishisato, S. (1980). *Analysis of Categorical Data. Dual Scaling and its Application*. University of Toronto Press.

O'Neill, M. E. (1978). Distributional Expansions for Canonical Correlations from Contingency Tables. *J. Roy. Statist. Soc., B*, Vol. 40. pp. 303–312.

O'Neill, M. E. (1981). A Note on the Canonical Correlations from Contingency Tables. *Aust. J. Statist.*, Vol. 23, pp. 58–66.

Parlett, B. N. (1980). *The Symmetric Eigenvalue Problem*. Prentice-Hall, Englewood Cliffs, N.J.

Pearson, K. (1901). On Lines and Planes of Closest Fit to Systems of Points in Space. *Phil. Mag.*, Vol. 2, No. 11, pp. 559–572.

Pillai, K. C. S. (1965). On the Distribution of the Largest Root of a Matrix in Multivariate Analysis. *Biometrika*, Vol. 52, pp. 405–414.

Pillai, K. C. S. (1967). Upper Percentage Point of the Largest Root of a Matrix in Multivariate Analysis. *Biometrika*, Vol. 54, pp. 189–194.

Pillai, K. C. S., and Chang, T. C. (1970). An Approximation to the C.D.F. of the Largest Root of a Covariance Matrix. *J. Inst. Statist. Math.*, pp. 115–124.

Prim, R. C. (1957). Shortest Connection Matrix Network and Some Generalizations. *Bell Syst. Tech. J.*, Vol. 36, pp. 1389–1401.

Rao, C. R. (1952). *Advanced Statistical Methods in Biometric Research*. Wiley, New York.

Rao, C. R. (1964). The Use and Interpretation of Principal Component Analysis in Applied Research. *Sankhya, A*, Vol. 26, 329–357.

Richardson, M.W., and Kuder, G. F. (1933). Making a Rating Scale that Measures. *Pers. J.*, Vol. 12, pp. 36–40.

Robbins, H., and Monro, S. (1951). A Stochastic Approximation Method. *Ann. Math. Statist.*, Vol. 22, pp. 400–407.

Romeder, J. M. (1973). *Méthodes et Programmes d'Analyse Discriminante*. Dunod, Paris.

Roy, S. N. (1939). *p*-Statistics or Some Generalisations of Analysis of Variance Appropriate to Multivariate Problems. *Sankhya*, Vol. 4, pp. 381–396.

Rutishauser, H. (1969). Solutions of Eigenvalue Problems with the LR Transformation. *Numer. Math.*, Vol. 13, pp. 4–13.

Rutishauser, H. (1970). Simultaneous Iteration Method for Symmetric Matrices. *Numer. Math.*, Vol. 16, pp. 205–223.

Schmetterer, L. (1958). Sur l'Itération Stochastique. *Colloq. Int. CNRS*, Vol. 87, pp. 55–63.

Schmetterer, L. (1969). Multidimensional Stochastic Approximation. In *Multivariate Analysis.*, Vol. 2 (P. R. Krishnaiah, Ed.). Academic Press, New York, pp. 443–460.

Schwartz, H. R., Rutishauser, H., and Stiefel, E. (1973). *Numerical Analysis of Symmetric Matrices*. Prentice-Hall, Englewood Cliffs, N.J., pp. 139–144.

Searle, S. R. (1982). *Matrix Algebra Useful for Statistics*. Wiley, New York.

Senne, K. D. (1974). Machine Independent Monte Carlo Evaluation of the Performance of Dynamic Stochastic Systems. *J. Stochastics*, Vol. 1, pp. 215–238.

Shepard, R. N. (1972). Introduction to Volume I. *In Multidimensional Scaling: Theory and Applications in the Behavioral Sciences* (R. N. Shepard, A. K. Romney, and S. B. Nerlove, Eds.). Seminar Press, New York.

Shepard, R. N. (1974). Representation of Structure in Similarity Data: Problems and Prospects. *Psychometrika*, Vol. 39, No. 4, pp. 373–421.

Sneath, P. H. A. (1957). Computer in Taxonomy. *J. Gen. Microbiol.*, Vol. 17, pp. 201–226.

Sneath, P. H. A., and Sokal, R. R. (1973). *Numerical Taxonomy*. Freeman, San Francisco.

Stewart, G. W. (1973). *Introduction to Matrix Computation*. Academic Press, New York.

Sugiyama, T. (1966). On the Distribution of the Largest Latent Root and the Corresponding Latent Vector for Principal Component Analysis. *Ann. Math. Statist.*, Vol. 37, pp. 995–1001.

Sylvester, J. J. (1889). Messenger of Mathematics (quoted by Eckart and Young, 1939), Vol. 19, No. 42.

Tabard, N. (1974). Besoins et Aspirations des Familles et des Jeunes. *Coll. Etudes C.A.F.*, No. 16.

Thom, R. (1974). Modèles Mathématiques de la Morphogénèse. *Coll.* 10/18, No. 887. Bourgois, Paris.

Thorndike, R. L. (1953). Who Belongs in the Family. *Psychometrika*, Vol. 18, pp. 267–276.

Tyler, D. E. (1981). Asymptotic Inference for Eigenvectors. *Ann. Statist.*, Vol. 9, pp. 725–736.

Wachter, K. W. (1978). The Strong Limits of Random Matrix Spectra for Sample Matrices of Independent Elements. *Ann. Prob.*, Vol. 6, No. 1, pp. 1–18.

Ward, Jr., J. H. (1963). Hierarchical Grouping to Optimise an Objective Function. *J. Amer. Statist. Assoc.*, Vol. 58, No. 301, pp. 236–244.

Watkins, D. S. (1982). Understanding the Q.R. Algorithm. *SIAM Rev.*, Vol. 24, pp. 427–440.

Wilkinson, J. H. (1965). *The Algebraic Eigenvalue Problem*. Clarendon Press, Oxford.

Wilkinson, J. H., and Reinsch, C. (1971) *Handbook for Automatic Computation. Vol. 2, Linear Algebra*, Springer-Verlag.

William, E. J. (1952). The Use of Score for the Analysis of Contingency Tables. *Biometrika*, Vol. 39, pp. 274–289.

Wong, M.A. (1982). A Hybrid Clustering Method for Identifying High-Density Clusters. *J. Amer. Statist. Assoc.*, Vol. 77, No. 380, pp. 841–847.

Index

Abbreviations used in the Index:

CA   Correspondence Analysis

GA   General Analysis

MCA   Multiple Correspondence Analysis

PCA   Principal Component Analysis